Progress in Nonlinear Differential Equations and Their Applications
Volume 51

Editor

Haim Brezis
Université Pierre et Marie Curie
Paris
and
Rutgers University
New Brunswick, N.J.

Editorial Board

Variational Methods for Discontinuous Structures

International Workshop at Villa Erba (Cernobbio), Italy, July 2001

Gianni dal Maso
Franco Tomarelli
Editors

Springer Basel AG

Editors' addresses:

Gianni dal Maso
Scuola Internazionale Superiore di Studi
Avanzati (SISSA)
Via Beirut 4
34014 Trieste
Italy
e-mail: dalmaso@sissa.it

Franco Tomarelli
Dipartimento di Matematica
"Francesco Brioschi"
Politecnico di Milano
Piazza Leonardo da Vinci 32
20133 Milano
Italy
e-mail: fratom@mate.polimi.it

2000 Mathematics Subject Classification 49-06; 65K10, 68U10, 74B99, 74C99, 74E25, 74N05, 82D40

A CIP catalogue record for this book is available from the Library of Congress, Washington D.C., USA

Bibliographic information published by Die Deutsche Bibliothek
Die Deutsche Bibliothek lists this publication in the Deutsche Nationalbibliografie; detailed bibliographic data is available in the Internet at <http://dnb.ddb.de>.

ISBN 978-3-0348-9470-8   ISBN 978-3-0348-8193-7 (eBook)
DOI 10.1007/978-3-0348-8193-7

© 2002 Springer Basel AG
Originally published by Birkhäuser Verlag in 2002
Printed on acid-free paper produced of chlorine-free pulp. TCF ∞

ISBN 978-3-0348-9470-8

9 8 7 6 5 4 3 2 1   www.birkhäuser-science.com

Contents

Preface

This volume contains the Proceedings of the International Workshop *Variational Methods For Discontinuous Structures*, which was jointly organized by the Dipartimento di Matematica Francesco Brioschi of Milano Politecnico and the International School for Advanced Studies (SISSA) of Trieste. The Conference took place at Villa Erba Antica (Cernobbio) on the Lago di Como on July 4–6, 2001.

In past years the calculus of variations faced mainly the study of continuous structures, say particularly problems with smooth solutions. One of the deepest and more delicate problems was the regularity of weak solutions. More recently, new sophisticated tools have been introduced in order to study discontinuities: in many variational problems solutions develop singularities, and sometimes the most interesting part of a solution is the singularity itself. The conference intended to focus on recent developments in this direction.

Some of the talks were devoted to differential or variational modelling of image segmentation, occlusion and textures synthesizing in image analysis, variational description of micro-magnetic materials, dimension reduction and structured deformations in elasticity and plasticity, phase transitions, irrigation and drainage, evolution of crystalline shapes; in most cases theoretical and numerical analysis of these models were provided.

Other talks were dedicated to specific problems of the calculus of variations: variational theory of weak or lower-dimensional structures, optimal transport problems with free Dirichlet regions, higher order variational problems, symmetrization in the BV framework.

This volume contains contributions by 12 of the 16 speakers invited to deliver lectures in the workshop. Most of the contributions present original results in fields which are rapidly evolving at present.

The meeting was successful in focusing on advanced results and techniques in pure and applied analysis of variational problems. The scientific level of the lectures was remarkable.

About seventy researchers in various fields (calculus of variations, geometric measure theory, materials science and computer science) attended the Conference. We take here the opportunity to thank them all for their contributions to the scientific discussion and the friendly atmosphere of the event.

The meeting would not have been possible without the support of several institutions and research projects. Here we wish to mention: national project *Calcolo delle Variazioni* co-financed by MIUR (2000), G.N.A.M.P.A. (INDAM), SIEMENS SpA Automation & Drives, Dipartimento di Matematica Francesco Brioschi (Milano Politecnico) and SISSA (Trieste).

Moreover we wish to thank the Centro di Cultura Scientifica *Alessandro Volta* (Como) for logistic support and for allowing to held the conference sessions in the beautiful Salone of Villa Erba.

We also thank for the excellent organizing work done by Ilaria Fragalà and Fabrizio Colombo, the efficient secretarial work done by Nadia Tansini, and for the hard typographical job done by Pamela Palazzini on the individual manuscripts.

Finally we thank all the lecturers, and we are especially grateful to those who contributed to this volume.

Levico, February 20, 2002   Gianni Dal Maso
                                                            Franco Tomarelli

Invited Lectures

EMILIO ACERBI, (Università di Parma)
A Model for Mixtures of Micromagnetic Materials allowing Existence and Regularity

GUY BOUCHITTÉ, (Université de Toulon et du Var)
Variational Theory of Weak Geometric Structures: the Measure Methods and its Applications

GIUSEPPE BUTTAZZO, (Università di Pisa)
Optimal Transportation Problems with Free Dirichlet Regions

GIANPIETRO DEL PIERO, (Università di Ferrara)
Interface Energies and Structured Deformations in Plasticity

IRENE FONSECA, (Carnegie-Mellon University of Pittsburgh)
Higher Order Variational Problems and Phase Transitions in Nonlinear Elasticity

NICOLA FUSCO, (Università di Firenze)
Symmetrization and Functionals Defined on BV

ROBERT V. KOHN, University of New York)
A Continuum Approach to the Structure of Twist Grain Boundaries

ANTONIO LEACI, (Università di Lecce)
Local Minimizers for a Free Gradient Discontinuity Problem in Image Segmentation

RICCARDO MARCH, (CNR, Roma)
Variational Properties of a Model for Image Segmentation with Overlapping Regions

JEAN MICHEL MOREL, (CMLA, France)
Irrigation

UMBERTO MOSCO, (Università di Roma)
  Effective Distances and Fractal Ramifications

DAVID OWEN, (Carnegie-Mellon University of Pittsburgh)
  Toward Field Theories for Bodies Undergoing Disarrangements

MAURIZIO PAOLINI, (Università Cattolica di Brescia)
  Unstable Crystalline Wulff Shapes in 3D

DANILO PERCIVALE, (Università di Genova)
  Dimension Reduction in Continuum Mechanics

JAYANT SHAH, (Northeastern University)
  Reaction-Diffusion Equations and Learning

LEV TRUSKINOVSKY, (University of Minneapolis)
  *Multiple Cracking: from Reinforced Concrete
  to Skeletal Muscles*

Progress in Nonlinear Differential Equations
and Their Applications, Vol. 51, 1–8

A Model for Mixtures
of Micromagnetic Materials
allowing Existence and Regularity

Emilio Acerbi

Abstract. In this paper a realistic model for mixtures of magnetic materials
is presented, and an application to an optimal design problem is also given.

1. Energies involved

In the Weiss-Landau-Lifschitz model of the micromagnetic theory (which we do
not want to discuss here in detail, referring the reader e.g. to [6]), the total energy
associated with a single crystal (or "grain") of a magnetic material is given by the
sum of several contributions. Every point (every atom in the real, discrete world)
in the body $\Omega \subset \mathbb{R}^3$ is magnetized, i.e., it generates a magnetic field, and we may
describe it with a vector field

$$\mathbf{m} : \Omega \to \mathbb{R}^3$$

called *magnetization*. Below a certain temperature characteristic of the material
(Curie temperature) the modulus of the magnetization is constant,

$$|\mathbf{m}| \equiv m^s \quad \text{in } \Omega ,$$

where the *magnetic saturation* intensity m^s is also a characteristic of the material.
Throughout the paper it is assumed that we are below the Curie temperature of
each material employed.

The first energy term is the *magnetic exchange energy,* generated by the wish
of adjacent points to share the same orientation of $\mathbf{m}$: since $|\mathbf{m}|$ is constant, the
exchange energy depends on the gradient matrix $\nabla \mathbf{m}$ through a four-indices tensor
A, also characteristic of the material, and is in general given by $\int_\Omega \langle A \nabla \mathbf{m}, \nabla \mathbf{m} \rangle \, dx$.
A very good approximation, which is commonly assumed, is that A is close to being
a multiple of the identity, thus we set

$$\text{Exch} = \int_\Omega a |\nabla \mathbf{m}|^2 \, dx ,$$

where the constant a is another characteristic of the material: when a is small the exchange energy is low, and we are in presence of a "soft" magnetic material as e.g. iron; when a is large, as in permanent magnets, the magnetic material is "hard".

A second character on the energy stage is the *anisotropy energy:* due to the structure of the crystal, there are some alignments of the magnetization **m** (the *easy axes*) which are preferred with respect to others: there is one direction (and its opposite) for uniaxial crystals, there are more for different symmetry groups. This is generally described as the integral of a polynomial in **m**, but we generalize it here as

$$\text{Anis} = \int_\Omega \phi(\mathbf{m})\, dx \ ,$$

where the continuous function

$$\phi : \partial B_{m^s} \to [0, +\infty[$$

depends on the material. In order to describe (in a qualitative way) the interplay of these two energies, rescale for a while the function ϕ by setting $\kappa = \max \phi$ and $\psi = \phi/\kappa$, so the two energy terms are

$$\text{Exch} + \text{Anis} = \int_\Omega [a|\nabla m|^2 + \kappa\psi(\mathbf{m})]\, dx \ .$$

Now consider a rod of uniaxial crystal, and assume that at the two ends the magnetization is along the easy axis, but points in the two opposite directions: the anisotropy energy wants **m** to stay preferably in these two directions, whereas the exchange energy term favours a slow transition. What conceivebly happens is that **m** will stay almost constant in two extremal portions of the rod, and there will be a transition layer between the two orientations; having a large and κ small leads to a slow transition, thus a thick interface, and having a small and κ large makes for sudden transition and narrow interface. According to one of the theories about how the transition occurs, it is easy to compute that the thickness of the interface is proportional to $\sqrt{a/\kappa}$ and that the total energy contribution of the layer is proportional to

$$\sqrt{a\kappa} \ \cdot \ \text{area of cross-section};$$

in other theories the numbers are different, but the area of the cross-section remains. The fact that this really happens may be seen in any magnetic material: the grain ends up divided into a lot of islands where the orientation of **m** is constant, the *magnetic domains,* separated by thin layers where all the transitions take place, the *Bloch walls* (according to other theories, there are other kinds of walls, such as e.g. *Néel walls,* to which some of our considerations apply only on a large-scale level). To get an idea of the dimensions involved, the diameter of magnetic domains is in the range of one tenth of a micron to millimeters, whereas the thickness of the walls is of about 10 to 100 atomic layers. We remark that many times, when doing explicit computations, the sum Exch + Anis is replaced by a constant (or a function of the magnetizations on the two sides) times the area of the middle section of the wall. We will later generalize the model in order to include all these variations.

Another, easy term in the expression of the energy is due to the *Zeeman field*, the influence of the external magnetic field $\mathbf{f}$ (to simplify the notation we dropped a few constants, such as $1/2$ in front of some energy terms):

$$\text{Ext} = -\int_\Omega \mathbf{f} \cdot \mathbf{m}\, dx\,,$$

of which little has to be said: $\mathbf{m}$ will try to align with $\mathbf{f}$.

The last term which is classically considered is the magnetostatic, or *dcmagnetizing energy*: $\mathbf{m}$ itself generates a magnetic field in the whole space, $\mathbf{h}[\mathbf{m}]$, which the magnetostatic equations give as

$$\begin{cases} \operatorname{curl} \mathbf{h} = \mathbf{0} & \text{in } \mathbb{R}^3 \\ \operatorname{div} (\mathbf{h} + \mathbf{m}\mathbf{1}_\Omega) = 0 & \text{in } \mathbb{R}^3. \end{cases} \tag{1}$$

These are to be interpreted, in a weak sense, as

$$\begin{cases} \mathbf{h} \in L^2(\mathbb{R}^3; \mathbb{R}^3)\,, \quad \operatorname{curl} \mathbf{h} = \mathbf{0}\,, \\ \int_{\mathbb{R}^3} \mathbf{h} \cdot \mathbf{v}\, dx = -\int_\Omega \mathbf{m} \cdot \mathbf{v}\, dx \quad \forall \mathbf{v} \in L^2(\mathbb{R}^3; \mathbb{R}^3) \text{ such that } \operatorname{curl} \mathbf{v} = \mathbf{0}\,; \end{cases}$$

then the demagnetizing energy (which is a nonlocal term) is given by

$$\text{Demag} = \int_{\mathbb{R}^3} |\mathbf{h}[\mathbf{m}]|^2\, dx\,.$$

Since (1) holds in $\mathbb{R}^3$, the generated magnetic field is zero if $\mathbf{m}$ is divergence-free and at the same time tangent to the boundary of Ω, whereas it is large if $\mathbf{m}$ has constant direction: thus the demagnetizing energy has large effects and heavily interferes with the exchange and anisotropy energies, which have opposite wishes regarding the alignment of $\mathbf{m}$. It is not the purpose of this paper to describe the interesting microstructure problems arising from this situation, for which we refer e.g. to [5].

The mapping $\mathbf{m} \mapsto \mathbf{h}[\mathbf{m}]$ has some interesting properties (see [4]): it is linear, continuous from $L^p(\Omega)$ to $L^p(\mathbb{R}^3)$ and

$$\int_{\mathbb{R}^3} |\mathbf{h}[\mathbf{m}]|^2\, dx = -\int_\Omega \mathbf{m} \cdot \mathbf{h}[\mathbf{m}]\, dx\,;$$

from this formula we deduce in particular that

$$\mathbf{m} \mapsto \int_{\mathbb{R}^3} |\mathbf{h}[\mathbf{m}]|^2\, dx \quad \text{is continuous from } L^2(\Omega; \mathbb{R}^3) \text{ to } \mathbb{R}\,.$$

Also, although the term is nonlocal, some local estimates may be recovered: if two magnetizations agree outside a ball, i.e. if $\mathbf{m} = \mathbf{m}'$ outside $B_\varrho \subset \Omega$, we have

$$\int_{\mathbb{R}^3} |\mathbf{h}[\mathbf{m}] - \mathbf{h}[\mathbf{m}']|^2\, dx = \int_{\mathbb{R}^3} |\mathbf{h}[\mathbf{m} - \mathbf{m}']|^2\, dx \le c\|\mathbf{m} - \mathbf{m}'\|_{L^2}^2 \le c\varrho^3$$

since $|\mathbf{m}| = |\mathbf{m}'| = m^s$; also, by the continuity of ϕ on the compact set ∂B_{m^s}

$$\int_\Omega |\phi(\mathbf{m}) - \phi(\mathbf{m}')|\, dx \leq c\varrho^3 :$$

both estimates are extremely useful when proving regularity.

The total energy associated with a magnetization $\mathbf{m}$ of a single crystal is given by the sum of the four terms we discussed, i.e.,

$$\begin{aligned} E(\mathbf{m}) &= \text{Exch} + \text{Anis} + \text{Ext} + \text{Demag} \\ &= \int_\Omega [a|\nabla\mathbf{m}|^2 + \phi(\mathbf{m}) - \mathbf{f}\cdot\mathbf{m}]\, dx + \int_{\mathbb{R}^3} |\mathbf{h}[\mathbf{m}]|^2\, dx . \end{aligned}$$

A self-evident defect of this model is that it only applies to a single crystal (it has other defects, the main of which is that it is valid only for approximately insulating materials, since conduction electrons play an important rôle in the magnetic theory of conductors).

2. Mixtures and their energy

We now turn to *mixtures* of magnetic materials: we begin by considering a body Ω made of two crystals Ω_1 and Ω_2 of different materials, separated by a smooth surface Σ; we stress the fact that by "different" we mean that the two grains may also be made of the same "stuff" but with crystallographic axes orientated in different directions. In general, as we have seen, a magnetic material is identified by just three quantities: the exchange constant a, the magnetic saturation m^s, and the anisotropy function ϕ, which contains all the necessary crystallographic information, thus we are confronted with two materials identified by the triples (a_1, m_1^s, ϕ_1) and (a_2, m_2^s, ϕ_2) respectively. The energy contribution of the magnetizations $\mathbf{m}_1$ and $\mathbf{m}_2$ of the two grains is thus given by

$$\begin{aligned} V(\mathbf{m}_1, \mathbf{m}_2) &= \int_{\Omega_1} [a_1|\nabla\mathbf{m}_1|^2 + \phi_1(\mathbf{m}_1) - \mathbf{f}\cdot\mathbf{m}_1]\, dx \\ &+ \int_{\Omega_2} [a_2|\nabla\mathbf{m}_2|^2 + \phi_2(\mathbf{m}_2) - \mathbf{f}\cdot\mathbf{m}_2]\, dx \\ &+ \int_{\mathbb{R}^3} |\mathbf{h}|^2\, dx , \end{aligned}$$

where $\mathbf{h} = \mathbf{h}[\mathbf{m}_1 \mathbf{1}_{\Omega_1} + \mathbf{m}_2 \mathbf{1}_{\Omega_2}]$.

In addition to this volume energy, an extra energy term on Σ has to be considered; the density of this surface energy is given by the sum of two terms: a positive constant $\alpha_{1,2}$ (depending on the two materials and due to chemical and electric disturbances in the lattice atoms), and a non-negative function β depending on the two materials, on the traces $\text{Tr}\, \mathbf{m}_1$ and $\text{Tr}\, \mathbf{m}_2$ of the magnetizations on the two sides of Σ, and possibly also on the normal vector ν to Σ. The function

$\beta_{1,2}(\operatorname{Tr} \mathbf{m}_1, \operatorname{Tr} \mathbf{m}_2, \nu)$ keeps track of the magnetic disturbances across Σ, and the surface term is

$$S(\mathbf{m}_1, \mathbf{m}_2) = \int_{\Sigma} [\alpha_{1,2} + \beta_{1,2}(\operatorname{Tr} \mathbf{m}_1, \operatorname{Tr} \mathbf{m}_2, \nu)] \, d\mathcal{H}^2 \, ,$$

so the total energy is

$$E(\mathbf{m}_1, \mathbf{m}_2) = V(\mathbf{m}_1, \mathbf{m}_2) + S(\mathbf{m}_1, \mathbf{m}_2) \, .$$

If the number of crystals grows, so does the shape of E; to keep it readable, we make for a while a simplification, which we will later drop to return to a general formulation: we assume $\beta = 0$ and $\alpha = 1$, so the surface term reduces to the area of the interface.

It is now easy to describe the case of K grains: we have K magnetic materials, each characterized by a triple (a_i, m_i^s, ϕ_i) and occupying an open subset Ω_i of Ω; the sets Ω_i are pairwise disjoint and their union is all of Ω up to a 2-dimensional set Σ; if we denote by $\mathbf{m}_i$ the magnetization in Ω_i, the energy is given by

$$\sum_{1}^{K} \int_{\Omega_i} [a_i |\nabla \mathbf{m}_i|^2 + \phi_i(\mathbf{m}_i) - \mathbf{f} \cdot \mathbf{m}_i] \, dx + \int_{\mathbb{R}^3} |\mathbf{h}[\textstyle\sum_{1}^{K} \mathbf{m}_i \mathbb{1}_{\Omega_i}]|^2 \, dx + \mathcal{H}^2(\Sigma) \, , \quad (2)$$

where we recall that

$$|\mathbf{m}_i| = m_i^s \quad \text{in } \Omega_i \, , \qquad \mathbf{m}_i \in W^{1,2}(\Omega_i; \mathbb{R}^3) \, . \quad (3)$$

3. A new form of the energy

We now turn our attention to an optimal design problem, the analysis of which will put under a new light the energy above, will force us to modify it once more, and will lead to a great simplification of the shape of the energy.

Fix a bounded open set $\Omega \subset \mathbb{R}^3$ and a vector field $\mathbf{f} \in L^2(\Omega; \mathbb{R}^3)$, and assume you have an infinite amount of each of K different magnetic materials, with which you want to fill Ω in order to minimize the energy (2): thus you want to find K pairs $(\Omega_i, \mathbf{m}_i)$ satisfying all the conditions above, and noticeably the constraints (3), which minimize the energy among all such sets of pairs. The trained eye will see at first glance that this problem is not well posed, because the energy is not semicontinuous: in particular, a magnetization $\mathbf{m}_i$ which is discontinuous along a surface $\varsigma \subset \Omega_i$ is not an admissible competitor, as it violates the Sobolev condition in (3), but it may be easily approached by a sequence of admissible functions with equibounded energy, simply by fattening ς into an open set ς' and adding this to Ω_j for some $j \neq i$ (extend $\mathbf{m}_j$ to ς' as a constant). This then leads to a finite relaxed energy for the discontinuous function we selected, although it is not in $W^{1,2}$; we remark that this is in some sense analogous to Gibbs' phenomenon in fluids (see [7]): if you want the magnetization to be discontinuous inside a crystal, break the crystal and coat the fracture faces with a different material.

The structure of the relaxed energy, which allows inner discontinuities but penalizes them, may be physically interpreted as keeping into account the possible

magnetic disarrangements ("magnetic cracks") inside a crystal, or as a simplification (which we already met) of the energy of a Bloch wall.

Since we have been forced to consider a relaxed energy which no longer forces
the magnetizations to be in $W^{1,2}$ inside each grain, but allows jumps, it is natural
to take as an ambient space that of special functions of bounded variation, SBV,
for whose definition and properties we refer to [2]. This setting allows us to write
the energy in a different form, but this is not simpler than before: indeed we cannot
charge all surface terms on the jump set of the overall magnetization, because some
parts of the surface may then be missing: to convince ourselves, recall that in two
adjacent grains Ω_i and Ω_j one may well have $m_i^s = m_j^s$, thus the magnetization
might have no jump across the interface, although some energy has to be taken
into account (due to the electric disturbances we mentioned). Also, it is impossible
to write the energy in a compact form.

We then rescale the magnetizations $\mathbf{m}_i$ in order to obtain an auxiliary magnetization field which will contain all the information, and which will allow us to
write the energy in an easy, implicit form. We set for $i = 1, \ldots, K$

$$\mathbf{u}_i = i \cdot \frac{\mathbf{m}_i}{m_i^s} , \qquad \mathbf{u} = \sum_1^K \mathbf{u}_i \mathbb{1}_{\Omega_i} ,$$

so that in particular

$$\mathbf{u} \in SBV(\Omega; \mathbb{R}^3) ,$$

$$\Omega_i = \{x : |\mathbf{u}| = i\}$$

and

$$\mathbb{1}_{\Omega_i}(x) = \left[2 - \left(|\mathbf{u}(x)| - (i-1)\right)^+\right]^+ := \mu_i\big(|\mathbf{u}(x)|\big) ,$$

$$\mathbf{m} = \mathbf{m}(\mathbf{u}) = \left(\sum_1^K \frac{m_i^s}{i} \mathbb{1}_{\Omega_i}\right)\mathbf{u} = \left(\sum_1^K \frac{m_i^s}{i} \mu_i(|\mathbf{u}|)\right)\mathbf{u} := \lambda(|\mathbf{u}|)\mathbf{u}$$

and the jump set $J_\mathbf{u}$ of $\mathbf{u}$ consists exactly of the union of both the interfaces
between grains and the inner magnetick cracks. We remark that given $\mathbf{u}$ one easily
deduces $\mathbf{m}$ and may also decide whether a jump of $\mathbf{u}$ represents an interface or an
inner crack: the former is also a jump of $|\mathbf{u}|$, the second is not.

We may now rescale all other factors: take any bounded, positive, continuous
function a satisfying

$$a : [0, +\infty[\to]0, +\infty[, \qquad a(i) = (m_i^s)^2\, a_i \quad \text{for } i = 1, \ldots, K$$

and we have

$$\text{Exch} = \sum_1^K \int_{\Omega_i} a_i |\nabla \mathbf{m}_i|^2\, dx = \int_\Omega a(|\mathbf{u}|)|\nabla \mathbf{u}|^2\, dx$$

(the extension of the function a outside the points $1, \ldots, K$ is not really necessary, but then this energy is defined on all SBV); analogously, take any bounded, non-negative, continuous function satisfying

$$\phi : \mathbb{R}^3 \to [0, +\infty[\, , \qquad \phi \big|_{\partial B_i}(\mathbf{z}) = \phi_i(m_i^s \mathbf{z}) \quad \text{for } i = 1, \ldots, K$$

and

$$\text{Anis} = \sum_1^K \int_{\Omega_i} \phi_i(\mathbf{m}_i) \, dx = \int_\Omega \phi(\mathbf{u}) \, dx \, .$$

Now, since the mapping $\mathbf{u} \mapsto \mathbf{m}(\mathbf{u}) = \lambda(|\mathbf{u}|)\mathbf{u}$ is continuous in every L^p, so is the mapping

$$\mathbf{u} \mapsto \mathbf{h}[\mathbf{m}(\mathbf{u})] := \tilde{\mathbf{h}}[\mathbf{u}]$$

(although it is no longer linear because λ is not), and we may write the whole energy as

$$\mathcal{E}(\mathbf{u}) = \int_\Omega \left[a(|\mathbf{u}|)|\nabla \mathbf{u}|^2 + \phi(\mathbf{u}) - \mathbf{f} \cdot \mathbf{m}(\mathbf{u}) - \tilde{\mathbf{h}}[\mathbf{u}] \cdot \mathbf{m}(\mathbf{u}) \right] dx + \int_{J_\mathbf{u}} \gamma(\mathbf{u}^+, \mathbf{u}^-, \nu) \, d\mathcal{H}^2 \, ,$$

where the function γ encompasses all the surface terms we met before. We make on γ the subadditivity assumptions that are customary to have subadditivity (see [2]), and due to the discussion above we also set

$$\gamma \geq \gamma_0 > 0 \, .$$

The optimal design problem then becomes

$$\min \left\{ \mathcal{E}(\mathbf{u}) : \mathbf{u} \in SBV(\Omega; \mathbb{R}^3), \, |\mathbf{u}| \in \{1, \ldots, K\} \text{ in } \Omega \right\} \, ,$$

a Mumford-Shah-type problem with a constraint on K different surfaces (see e.g. [3] for a single constraint in a simpler setting).

In a forthcoming paper [1] it is proved that $\mathcal{E}$ is semicontinuous and the minimum exists (also in the fixed-volume-fraction case), and a regularity theorem for the solution and the interfaces is given.

Acknowledgement. This paper was conceived during a stay at Carnegie Mellon University, and written during a stay as a guest of the Sonderforschungsbereich 256 of the University of Bonn; the authors warmly thanks both institutions, as well as MIUR and GNAMPA who provided partial financing through Cofin/2000 funds and the research project "Modelli variazionali sotto ipotesi non standard".

References

[1] E. Acerbi, I. Fonseca, G. Mingione: *paper in preparation.*

[2] L. Ambrosio, N. Fusco, D. Pallara: *Functions of bounded variation and free discontinuity problems,* Clarendon Press, Oxford University Press, New York, 2000.

[3] M. Carriero, A. Leaci: S^k*-valued maps minimizing the L^p norm of the gradient with free discontinuities,* Ann. Scuola Norm. Sup. Pisa Cl. Sci. (4) **18** (1991), 321–352.

[4] A. DeSimone: *Energy minimisers for large ferromagnetic bodies*, Arch. Rational Mech. Anal. **125** (1993), 99–143.

[5] A. DeSimone, R.V. Kohn, S. Müller, F. Otto: *Magnetic microstructures – a paradigm of multiscale problems*, ICIAM 99 (Edinburgh), 175–190, Oxford Univ. Press, Oxford, 2000.

[6] A. Hubert, R. Schäfer: *Magnetic domains*, Springer Verlag, Berlin, Heidelberg, New York, 1998.

[7] L. Modica: *Gradient theory of phase transitions with boundary contact energy*, Ann. Inst. H. Poincaré Anal. Non Linéaire **4** (1987), 487–512.

Emilio Acerbi
Department of Mathematics,
University of Parma,
V. D'Azeglio, 85/a
I–43100 Parma, Italy
E-mail address: emilio.acerbi@unipr.it

Progress in Nonlinear Differential Equations
and Their Applications, Vol. 51, 9–17

Variational Properties of a Model for Image Segmentation with Overlapping Regions

Giovanni Bellettini and Riccardo March

Abstract. We introduce a functional for image segmentation which takes into account the occlusions between objects in the image which are located at different depths in space. By minimizing the functional, we try to reconstruct both a piecewise smooth approximation of the input image g and the contours of the objects together with their hidden portions. Some variational properties of the involved functionals are then studied.

1. Introduction of the model

In recent years several variational models related to the image segmentation problem have been proposed [10, 11]. In the present paper we introduce a functional inspired to the theory of segmentation with depth proposed by Nitzberg and Mumford [12] and by Nitzberg, Mumford and Shiota [13]; by minimizing the functional, our aim is to reconstruct both a piecewise smooth approximation of the input image g and the contours of the objects together with their hidden portions.

Let us precisely introduce the model. Let $\varphi \in BV(\mathbb{R}^2; \mathbb{Z})$; then φ can be written as

$$\varphi = \sum_{i \in I} \alpha_i \chi_{E_i}, \qquad \text{with } \alpha_i \in \mathbb{Z} \setminus \{0\} \text{ for any } i \in I,$$

where $\{E_i\}_i$ is a sequence of *possibly overlapping* finite perimeter sets, $E_i \neq E_j$ for any $i, j \in I$, $i \neq j$, $I \subseteq \mathbb{N}$ is at most countable, and χ_{E_i} is the characteristic function of E_i. We call the double sequence $\{\alpha_i, E_i\}_{i \in I}$ a *representation* of φ.

We define the functional $\mathcal{E} : BV(\mathbb{R}^2; \mathbb{Z}) \to [0, +\infty]$ as follows:

$$\mathcal{E}(\varphi) := \inf \left\{ \sum_{i \in I} |\alpha_i| \overline{\mathcal{F}}(E_i) : \varphi = \sum_{i \in I} \alpha_i \chi_{E_i} \right\}. \tag{1}$$

The definition of the functional $\mathcal{E}$ is based on another functional $\mathcal{F}$, which can be viewed as the core of the model, and was firstly introduced in [12]. Precisely, let $\psi : \mathbb{R} \to [0, +\infty[$ be the convex $\mathcal{C}^1$ function defined by

$$\psi(\xi) = \psi(|\xi|) := \begin{cases} c_1 \xi^2 & \text{if } |\xi| < T \\ c_2 |\xi| - c_3 & \text{if } |\xi| \geq T, \end{cases} \tag{2}$$

with $T > 0$, $c_1 > 0$, $c_1 T^2 = c_2 T - c_3$, $c_2 = 2c_1 T$. Denoting by $\mathcal{C}_b^2(\mathbb{R}^2)$ the class of all bounded open subsets of $\mathbb{R}^2$ of class $\mathcal{C}^2$, we define the map $\mathcal{F} : L^1(\mathbb{R}^2) \to [0, +\infty]$ by

$$
\mathcal{F}(v) := \begin{cases} \displaystyle\int_{\partial E} [1 + \psi(\kappa)] \, d\mathcal{H}^1 & \text{if } v = \chi_E, \ E \in \mathcal{C}_b^2(\mathbb{R}^2) \\[2mm] +\infty & \text{elsewhere,} \end{cases} \tag{3}
$$

κ being the curvature of ∂E and $\mathcal{H}^1$ the one-dimensional Hausdorff measure.

In the following the quantity $\mathcal{F}(\chi_E)$ will be denoted by $\mathcal{F}(E)$, thus emphasizing the dependence of $\mathcal{F}$ on the set E rather than on its boundary ∂E.

The functional $\overline{\mathcal{F}}$ appearing in the definition of $\mathcal{E}$ is then the $L^1(\mathbb{R}^2)$-relaxed functional of $\mathcal{F}$, i.e.,

$$
\overline{\mathcal{F}}(E) := \inf \left\{ \liminf_{h \to +\infty} \mathcal{F}(E^h) : \{E^h\}_h \subseteq \mathcal{C}_b^2(\mathbb{R}^2), E^h \to E \text{ in } L^1(\mathbb{R}^2) \right\}.
$$

In [3] it has been proved that $\mathcal{F} = \overline{\mathcal{F}}$ on $\mathcal{C}_b^2(\mathbb{R}^2)$.

To better understand the meaning of the functional $\mathcal{E}$, and in particular the meaning of the energy of an optimal representation of φ, let us consider the following example. Let $\varphi \in BV(\mathbb{R}^2; \{0, 1, 2\})$ be the function whose values are represented in the two overlapping circles of Figure 1; if we indicated by C_l (resp. C_r) the interior of the left (resp. right) circle, then φ has value 1 on $C_l \setminus C_r$ and on $C_r \setminus C_l$, while $\varphi = 2$ on $C_l \cap C_r$, and $\varphi = 0$ out of the two circles. The function φ has different representations $\sum_i \alpha_i \chi_{E_i}$, each of which has a corresponding associated energy $\sum_i |\alpha_i| \overline{\mathcal{F}}(E_i)$. Figure 1 shows two different representations of φ; the left one consists of writing φ as

$$
\varphi = \chi_{C_l} + \chi_{C_r},
$$

whose associated energy is given by

$$
\mathcal{F}(C_l) + \mathcal{F}(C_r) = 2 \int_{\partial C_l} [1 + \psi(\kappa)] \, d\mathcal{H}^1. \tag{4}
$$

On the other hand, φ admits also the nonsmooth representation

$$
\varphi = \chi_{C_l \cup C_r} + \chi_{C_l \cap C_r}
$$

which is depicted at the right in Figure 1. The energy of this representation is

$$
\overline{\mathcal{F}}(C_l \cup C_r) + \overline{\mathcal{F}}(C_l \cap C_r), \tag{5}
$$

notice the presence of $\overline{\mathcal{F}}$ instead of $\mathcal{F}$, being $C_l \cup C_r$ and $C_l \cap C_r$ nonsmooth sets. Now, it is not difficult to verify that the energy in (4) is strictly smaller than the energy in (5); this is essentially due to the fact that, when computing $\overline{\mathcal{F}}(E)$ on a nonsmooth set E, corner points of ∂E are source either of concentration of energy or of a contribution of energy due to some hidden curve (see [3]). In this example the representation of φ with minimal energy corresponds to the two smooth partially overlapping circles C_r and C_l, and the value of the energy is the same as the one considered in [13].

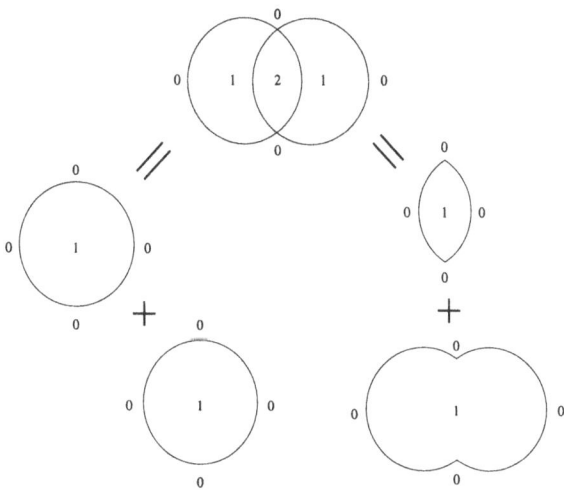

FIGURE 1. Representations of two circles with an occlusion

In the sequel, we denote by $\mathcal{R}(\varphi)$ the set of all representations $\{\alpha_i, E_i\}_{i \in I}$ of φ such that $\overline{\mathcal{F}}(E_i) < +\infty$ for any $i \in I$.

Let us now turn to the complete definition of the functionals of the model.

Let $\Omega \subset \mathbb{R}^2$ be a bounded open set, and let $BV(\Omega)$ (resp. $SBV(\Omega)$) be the space of functions (resp. special functions) of bounded variation in Ω [1, 8]. Let $g \in L^2(\Omega)$ be the input image; we define the functional $\mathcal{M} : SBV(\Omega) \to [0, +\infty]$ as the usual Mumford-Shah functional in its weak form, i.e.,

$$\mathcal{M}(u) := \int_\Omega |\nabla u|^2 \, dx + \mathcal{H}^1(J_u) + \int_\Omega |u - g|^2 \, dx,$$

J_u being the jump set of u. We finally let $\mathcal{G} : SBV(\Omega) \times BV(\mathbb{R}^2; \mathbb{Z}) \to [0, +\infty]$ be defined as

$$\mathcal{G}(u, \varphi) := \begin{cases} \mathcal{M}(u) + \mathcal{E}(\varphi) & \text{if } J_u \subseteq J_\varphi \cap \Omega, \\ \\ +\infty & \text{elsewhere,} \end{cases} \tag{6}$$

and we set $\mathcal{D}(\mathcal{G}) := \{(u, \varphi) \in SBV(\Omega) \times BV(\mathbb{R}^2; \mathbb{Z}) : J_u \subseteq J_\varphi \cap \Omega\}$.

The functional $\mathcal{G}$ is the energy describing our model. The family $\{E_i\}_i$ represents a collection of objects in the input image g, the boundaries ∂E_i represent the contours of the objects, and the jump set J_u represents the portions of such contours which are visible in the image g. By minimizing the functional $\mathcal{G}$, we try to reconstruct both a piecewise smooth (i.e., denoised) approximation u of the image g and the contours of the objects ∂E_i. Possibly overlapping sets E_i are considered in order to model the occlusions between objects located at different depths in space. Then the model yields also a reconstruction of the hidden portions of the contours. Finally, we remind that the coefficients α_i are not related to the grey level of the image g.

Even if related, our model is not immediately comparable with the model in [13]: our geometric part $\mathcal{E}$ is, in some sense, simpler since it is defined on functions φ only, instead that on families of (possibly overlapping) sets partitioning the image domain. This simplification not always allows to reconstruct the spatial order (i.e., relative depth) of the different objects, which instead is one of the main features of the model of [13]. We can, however, always reconstruct the hidden curves (which are contained in J_φ), as the authors in [13]. Notice that we can insert in our model, by means of the functional $\mathcal{M}$, a not necessarily piecewise constant grey level u (on this subject some useful comments can be found also in Chapter 6 of [13]). The most striking difference between the two models probably consists in the lower order term, which in the present paper reads simply as $\int_\Omega |u - g|^2\, dx$. Finally, it must be noted that at least the functional $\mathcal{E}$ can be approximated in the variational sense of Γ-convergence with a sequence of simpler functionals, thus making possible a direct implementation on a computer (and a rather automatic way to reconstruct the hidden lines).

We conclude this short introduction by recalling that complete proofs of the results presented here will be published elsewhere [5].

2. Variational properties of the functional $\mathcal{E}$

We begin with a proposition which shows that $\mathcal{E}(\chi_E)$ coincides with $\overline{\mathcal{F}}(E)$ if E is a smooth set. In general however $\mathcal{E}(\chi_E)$ is smaller than $\overline{\mathcal{F}}(E)$ for nonsmooth sets; in particular, $\mathcal{E}$ does not extend the functional $\mathcal{F}$ on $\mathbb{Z}$-valued BV-functions.

Proposition 2.1. *Let E be such that $\overline{\mathcal{F}}(E) < +\infty$. Then $\mathcal{E}(\chi_E) \leq \overline{\mathcal{F}}(E)$. Moreover, if $E \in \mathcal{C}_b^2(\mathbb{R}^2)$ then $\mathcal{E}(\chi_E) = \overline{\mathcal{F}}(E)$.*

It can be proved that if $\varphi \in BV(\mathbb{R}^2; \mathbb{Z})$ is such that $\mathcal{E}(\varphi) < +\infty$, then φ admits a representation in $\mathcal{R}(\varphi)$ with a finite number of sets; moreover its jump set J_φ is contained in the union of a finite number of closed regular curves of class $W^{1,1}$ whose derivative is of class BV.

We now give a lower semicontinuity result for the functional $\mathcal{E}$.

Theorem 2.2. *Let $\{\varphi_h\}_h \subset BV(\mathbb{R}^2; \mathbb{Z})$ be a sequence converging to $\varphi \in BV(\mathbb{R}^2; \mathbb{Z})$ in $L^1_{\mathrm{loc}}(\mathbb{R}^2)$. Then*

$$\mathcal{E}(\varphi) \leq \liminf_{h \to +\infty} \mathcal{E}(\varphi_h).$$

We now give a compactness result for the functional $\mathcal{E}$; in order to give a complete statement, we need some preliminaries on systems of curves (see [3]).

We call regular curve any function $\gamma : [0,1] \to \mathbb{R}^2$ such that $|\dot\gamma| \neq 0$ in $[0,1]$. By $(\gamma) = \{\gamma(t) : t \in [0,1]\}$ we denote the trace of γ and by $l(\gamma)$ its length. If $x \in \mathbb{R}^2 \setminus (\gamma)$, $I(\gamma, x)$ denotes the index of x with respect to γ [6].

We call system of curves a finite family $\Gamma = \{\gamma^1, \ldots, \gamma^m\}$ of closed regular curves of class $W^{1,1}$ such that $|\dot\gamma^i|$ is constant almost everywhere on $[0,1]$ for any $i = 1, \ldots, m$. The trace (Γ) of a system of curves Γ is defined as $\cup_{i=1}^m (\gamma^i)$, and the length $l(\Gamma)$ of Γ as $\sum_{i=1}^{m} l(\gamma^i)$. If $x \in \mathbb{R}^2 \setminus (\Gamma)$, we define the index $I(\Gamma, x)$ of x with respect to Γ as $\sum_{i=1}^{m} I(\gamma^i, x)$.

We say that a system of curves $\Gamma = \{\gamma^1, \ldots, \gamma^m\}$ is of class $\mathcal{B}$, and we write $\Gamma \in \mathcal{B}$, if

$$\gamma_i \in W^{1,1}([0,1]; \mathbb{R}^2), \qquad \dot{\gamma}^i \in BV([0,1]; \mathbb{R}^2), \qquad \forall i = 1, \ldots, m.$$

We say that a sequence $\{\Gamma_h\}_h$ of systems of curves of class $\mathcal{B}$ is weakly convergent to a family of curves $\Gamma = \{\gamma^1, \ldots, \gamma^m\}$, if each of the systems Γ_h consists of a number m of curves $\{\gamma_h^1, \ldots, \gamma_h^m\}$, m independent of h, and $\gamma_h^i \to \gamma^i$ uniformly, $\dot{\gamma}_h^i \rightharpoonup \dot{\gamma}^i$ weakly in BV as $h \to +\infty$, for any $i = 1, \ldots, m$.

We say that Γ is a limit system of curves if Γ is the weak limit of a sequence of oriented parametrizations of bounded open sets of class $\mathcal{C}^2$ (see [2, 3]); then it can be proved that $I(\Gamma, x) \in \{0, 1\}$ for any $x \in \mathbb{R}^2 \setminus (\Gamma)$.

We say that a sequence $\{\Gamma_h\}_h$ of systems of curves of class $\mathcal{B}$ is weakly convergent to a family of curves $\Gamma = \{\gamma^1, \ldots, \gamma^m\}$ up to a finite set of points P, if the following conditions are satisfied:

(i) each of the systems Γ_h contains a number m of curves $\{\gamma_h^1, \ldots, \gamma_h^m\}$, m independent of h, such that the sequence of systems $\{\gamma_h^1, \ldots, \gamma_h^m\}$ converges weakly to Γ as $h \to +\infty$;

(ii) the maximum distance of the trace of the remaining curves of Γ_h (i.e., $(\Gamma_h \setminus \{\gamma_h^1, \ldots, \gamma_h^m\})$) from the set P goes to zero as $h \to +\infty$.

Finally, we denote by $|\cdot|$ the Lebesgue measure in $\mathbb{R}^2$ and by $B_r(x)$ the open ball of radius r centered at x.

We are now in a position to state the compactness theorem for the functional $\mathcal{E}$.

Theorem 2.3. *Let $\{\varphi_h\}_h \subset BV(\mathbb{R}^2; \mathbb{Z})$ be a sequence such that*

$$\sup_{h \in \mathbb{N}} \mathcal{E}(\varphi_h) < +\infty.$$

Then there exist a function $\varphi \in BV(\mathbb{R}^2; \mathbb{Z})$ and a subsequence $\{\varphi_{h_k}\}_k$ such that:

(a) *φ admits a finite representation $\{\alpha_i, E_i\}_{i \in I} \in \mathcal{R}(\varphi)$ and $\{\varphi_{h_k}\}_k$ converges to φ in $L^1_{\mathrm{loc}}(\mathbb{R}^2)$;*

(b) *there exist limit systems of curves $\Gamma_i \in \mathcal{B}$, $i \in I$, with*

$$\{x \in \mathbb{R}^2 : \exists r > 0 : |B_r(x) \setminus E_i| = 0\} = \mathrm{int}(A_{\Gamma_i} \cup (\Gamma_i)), \qquad i \in I, \qquad (7)$$

where $A_{\Gamma_i} := \{x \in \mathbb{R}^2 \setminus (\Gamma_i) : I(\Gamma_i, x) = 1\}$, and there exists a sequence $\{\Gamma^k\}_k \subset \mathcal{B}$ of systems of curves converging weakly to Γ, where $(\Gamma) := \cup_{i \in I}(\Gamma_i)$, up to a finite set of points P, such that

$$\forall r > 0 \;\; \exists k_r \in \mathbb{N} : \; J_{\varphi_{h_k}} \cap B_r(0) \subseteq (\Gamma^k) \qquad \forall k \geq k_r,$$

$$J_\varphi \subseteq (\Gamma). \qquad (8)$$

We conclude this section with a theorem of approximation of the functional $\mathcal{E}$ via Γ-convergence, see [7, 9].

For any $v \in C_c^\infty(\mathbb{R}^2)$ and $h \in \mathbb{N}$ we define the non-negative Radon measures μ_h on $\mathbb{R}^2$ by

$$d\mu_h(x) := \left[h^{-1}|\nabla v|^2 + hW(v) \right] dx,$$

where $W : \mathbb{R} \to [0, +\infty[$ is defined by $W(t) := t^2(1-t)^2$. We set $\lambda := 2 \int_0^1 \sqrt{W(t)} dt$.

For any $h \in \mathbb{N}$ let $\mathcal{F}_h : L^1(\mathbb{R}^2) \to [0, +\infty]$ be the map defined by

$$\mathcal{F}_h(v) := \begin{cases} \displaystyle\int_{\mathbb{R}^2 \setminus \{|\nabla v|=0\}} \left[1 + \psi\left(\mathrm{div}\left(\frac{\nabla v}{|\nabla v|} \right) \right) \right] d\mu_h(x) & \text{if } v \in C_c^\infty(\mathbb{R}^2), \\ \\ +\infty & \text{elsewhere.} \end{cases}$$

In [4] it has been proved that

$$\Gamma - \lim_{h \to +\infty} \mathcal{F}_h = \lambda \overline{\mathcal{F}} \qquad \text{on } L^1(\mathbb{R}^2).$$

For any $h \in \mathbb{N}$ let $\mathcal{E}_h : L^1(\mathbb{R}^2) \to [0, +\infty]$ be the map defined by

$$\mathcal{E}_h(v) := \inf \left\{ \sum_{i \in I} |\beta_i| \mathcal{F}_h(v_i) : v = \sum_{i \in I} \beta_i v_i, I \subseteq \mathbb{N}, \ v_i \in C_c^\infty(\mathbb{R}^2), \beta_i \in \mathbb{Z} \ \forall i \in I \right\},$$

if $v \in C_c^\infty(\mathbb{R}^2)$, and $\mathcal{E}_h(v) = +\infty$ elsewhere in $L^1(\mathbb{R}^2)$. Then the following theorem holds.

Theorem 2.4. *We have*

$$\Gamma - \lim_{h \to +\infty} \mathcal{E}_h = \lambda \mathcal{E} \qquad \text{on } L^1(\mathbb{R}^2).$$

Theorem 2.4 is a first step toward the approximation of the full functional $\mathcal{G}$, which is a problem deserving further investigation.

3. Relaxation properties of the functional $\mathcal{G}$

In this section we need functionals defined on systems of curves of class $\mathcal{B}$.

For any $f \in BV([0, 1])$, $f^-(x)$ and $f^+(x)$ denote the approximate lower and upper limit of f at the point x, respectively, $\dot{f}$ denotes the distributional derivative of f, $|\dot{f}|$ denotes the total variation of $\dot{f}$, $\dot{f}^a$ denotes the density of the absolutely continuous part of $\dot{f}$ with respect to the Lebesgue measure, and $\dot{f}^s$ denotes the singular part of $\dot{f}$.

Let $\gamma : [0, 1] \to \mathbb{R}^2$ be a curve of class $\mathcal{B}$ such that $|\dot{\gamma}| = R$ almost everywhere on $[0, 1]$, for a suitable $R > 0$. Then it can be proved (see [3], Lemma 3.1) that there exists a function $\omega : [0, 1] \to \mathbb{R}$, called an argument of $\dot{\gamma}$, such that $\omega \in BV(]0, 1[)$, $\dot{\gamma}(t) = (R\cos\omega(t), R\sin\omega(t))$ a.e. in $]0, 1[$, $J_\omega = J_{\dot{\gamma}}$, and $-\pi < \omega^+(t) - \omega^-(t) \leq \pi$ for every $t \in]0, 1[$.

We now introduce energy functionals defined on systems of curves of class $\mathcal{B}$ using the argument functions. Let $\gamma \in \mathcal{B}$ and let ω be an argument of $\dot\gamma$ on $[0,1]$. We define the functional

$$\mathcal{K}(\gamma) := l(\gamma)\int_0^1 \psi(l(\gamma)^{-1}\dot\omega^a)dt + c_2 \sum_{t\in[0,1]\cap J_\omega} \left|\omega^+(t) - \omega^-(t)\right| + c_2\left|\dot\omega^s\right|(]0,1[\setminus J_\omega).$$

The functional $\mathcal{K}(\gamma)$ is decomposed according to the contributions of the absolutely continuous part of $\dot\omega$, of the discontinuity set of ω, and of the Cantor part of $\dot\omega$. The value of the functional $\mathcal{K}$ does not depend on the choice of the argument ω.

Let $\Gamma = \{\gamma^1,\ldots,\gamma^m\}$ be a system of curves of class $\mathcal{B}$ and let ω_i be an argument of $\dot\gamma^i$ on $[0,1]$ for any $i = 1,\ldots,m$. We define

$$\mathcal{F}(\Gamma) := l(\Gamma) + \mathcal{K}(\Gamma), \qquad \mathcal{K}(\Gamma) := \sum_{i=1}^m \mathcal{K}(\gamma^i).$$

We are now in a position to state some results on the functional $\mathcal{G}$: for convenience, we relax $\mathcal{G}$ by means of the functional $\overline{\mathcal{G}} : SBV(\Omega) \times BV(\mathbb{R}^2;\mathbb{Z}) \to [0,+\infty]$ defined as

$$\overline{\mathcal{G}}(u,\varphi) := \inf\left\{\liminf_{h\to+\infty} \mathcal{G}(u_h,\varphi_h) : (u_h,\varphi_h) \to (u,\varphi) \text{ in } L^1(\Omega) \times L^1_{\mathrm{loc}}(\mathbb{R}^2)\right\}.$$

Remark 3.1. *If $\mathcal{E}(\varphi) < +\infty$, then φ admits a finite representation $\{\alpha_i, E_i\}_{i\in I} \in \mathcal{R}(\varphi)$ and there exist Γ_i and Γ as in (b) of Theorem 2.3 satisfying (7) and (8).*

Our results on the functional $\mathcal{G}$ read as follows.

Theorem 3.2. *Let $(u,\varphi) \in SBV(\Omega) \times BV(\mathbb{R}^2;\mathbb{Z})$. If $(u,\varphi) \in \mathcal{D}(\mathcal{G})$ then $\overline{\mathcal{G}}(u,\varphi) = \mathcal{G}(u,\varphi)$. If $\overline{\mathcal{G}}(u,\varphi) < +\infty$, then*

$$\overline{\mathcal{G}}(u,\varphi) \geq \mathcal{M}(u) + \mathcal{E}(\varphi),$$

and

$$u \in W^{1,2}(\Omega \setminus (\Gamma)), \tag{9}$$

where Γ is a system of curves as in Remark 3.1, $(\Gamma) = \cup_{i\in I}(\Gamma_i)$. Moreover

$$\overline{\mathcal{G}}(u,\varphi) \geq \mathcal{M}(u) + \sum_{i\in I}|\alpha_i|\mathcal{F}(\Gamma_i).$$

Finally, the following estimates hold.

Proposition 3.3. *The following two assertions hold.*

(a) *Let $(u,\varphi) \in SBV(\Omega) \times BV(\mathbb{R}^2;\mathbb{Z})$ be such that $\overline{\mathcal{G}}(u,\varphi) < +\infty$. Then*

$$\overline{\mathcal{G}}(u,\varphi) \geq \mathcal{M}(u) + \mathcal{A}(u,\varphi), \qquad \mathcal{A}(u,\varphi) := \inf_{\mathcal{R}(\varphi)} \inf_{\Gamma} \sum_{i\in I}|\alpha_i|\mathcal{F}(\Gamma_i),$$

where the infimum in the definition of $\mathcal{A}$ is over all systems of curves $\Gamma \in \mathcal{B}$ satisfying (7), (9) and $(\Gamma) = \cup_{i\in I}(\Gamma_i)$.

(b) *Let $\varphi \in BV(\mathbb{R}^2; \mathbb{Z})$ be such that $\mathcal{E}(\varphi) < +\infty$, and let $u \in W^{1,2}(\Omega \backslash \overline{J}_u)$. Let the sets $J_u \backslash \overline{J}_\varphi$ and $J_u \cap \overline{J}_\varphi$ have a finite number of connected components. Moreover assume that there exists a finite set of points F such that the set $(J_u \backslash \overline{J}_\varphi) \backslash F$ can be written, locally, as the graph of a function of class $\mathcal{C}^2$ up to its closure.*

Then there exists a system of curves $\Lambda \in \mathcal{B}$ such that $J_u \backslash \overline{J}_\varphi \subseteq (\Lambda)$, $|A_\Lambda| = 0$ with $A_\Lambda := \{x \in \mathbb{R}^2 \backslash (\Lambda) : I(\Lambda, x) = 1\}$, and

$$\overline{\mathcal{G}}(u, \varphi) \leq \mathcal{M}(u) + \mathcal{E}(\varphi) + \mathcal{F}(\Lambda).$$

References

[1] L. Ambrosio, *A compactness theorem for a new class of functions of bounded variation*, Boll. Un. Mat. Ital., **3-B** (1989), 857–881.

[2] G. Bellettini, G. Dal Maso and M. Paolini, *Semicontinuity and relaxation properties of a curvature depending functional in 2D*, Ann. Scuola Norm. Sup. Pisa Cl. Sci. **20** (1993), 247–299.

[3] G. Bellettini and M. Paolini, *Variational properties of an image segmentation functional depending on contours curvature*, Adv. Math. Sci. Appl., **5** (1995), 681–715.

[4] G. Bellettini, *Variational approximation of functionals with curvatures and related properties*, J. Conv. Anal., **4** (1997), 91–108.

[5] G. Bellettini and R. March, *An image segmentation variational model with free discontinuities and contour curvature*, paper in preparation.

[6] M. Berger and B. Gostiaux, *Differential Geometry: Manifolds, Curves and Surfaces*, Springer-Verlag, New York, **1988**.

[7] G. Dal Maso, *An Introduction to Γ-Convergence*, Birkhäuser, Boston, **1993**.

[8] E. De Giorgi and L. Ambrosio, *Un nuovo funzionale del calcolo delle variazioni*, Atti Accad. Naz. Lincei Cl. Sci. Fis. Mat. Natur. Rend. Lincei (8) Mat. Appl., **82** (1988), 199–210.

[9] E. De Giorgi and T. Franzoni, *Su un tipo di convergenza variazionale*, Atti Accad. Naz. Lincei Rend. Cl. Sci. Fis. Mat. Natur., **58** (1975), 842–850.

[10] J.M. Morel and S. Solimini, *Variational Methods in Image Segmentation*, Birkhäuser, Basel, **1995**.

[11] D. Mumford and J. Shah, *Optimal approximations by piecewise smooth functions and associated variational problems*, Comm. Pure Appl. Math., **42** (1989), 577–685.

[12] M. Nitzberg and D. Mumford, *The 2.1-D sketch*, in Proc. of the Third International Conference on Computer Vision, Osaka (1990).

[13] M. Nitzberg, D. Mumford and T. Shiota, *Filtering, Segmentation and Depth*, Lecture Notes in Computer Science **662**, Springer-Verlag, Berlin, 1993.

Giovanni Bellettini,
Dipartimento di Matematica,
Università di Roma "Tor Vergata",
Via della Ricerca Scientifica,
00133 Roma, Italy
E-mail address: belletti@mat.uniroma2.it

Riccardo March,
Istituto per le Applicazioni del Calcolo, CNR,
Viale del Policlinico 137,
00161 Roma, Italy
E-mail address: march@iac.rm.cnr.it

Progress in Nonlinear Differential Equations
and Their Applications, Vol. 51, 19–40

Variational Theory of Weak Geometric Structures: The Measure Method and Its Applications

Guy Bouchitté and Ilaria Fragalà

Abstract. We present a survey of results for different kinds of variational problems where weak geometric structures intervene as a common feature. A unified method is adopted, which has been developed in several papers during recent years; it is based on the use of measures, fitted out with suitable tangential properties and functional spaces. Some of the results are new, and their proof will be given in a forthcoming paper.

1. Introduction

Several theories have been developed in last decades in order to give a weak description of sets. In particular, the theory of varifolds due to Allard [1], and the theory of currents introduced by Federer and Fleming [22] date both around the sixties, and have by now found a systematic treatment (see the books [21], [25], [30]). In these frameworks, suitable compactness theorems hold, which are useful to prove existence results by the direct method of the Calculus of Variations, for instance concerning minimal surfaces. Nevertheless, the application of these theories to variational or optimization problems is still rather restricted. Besides technical difficulties, in our opinion the following two reasons may be pointed out.

First, it is easy to convince that often in optimization problems the dimension of the optimal structure may have a Hausdorff dimension *a priori* unknown, or even may be a multijunction, made by parts of different Hausdorff dimensions. In spite, the subsets of $\mathbb{R}^n$ under consideration in the settings of varifolds and currents must have a *fixed, integer* dimension $k \leq n$. Indeed, such theories are founded upon a notion of approximate tangent space, defined by blow-up, which makes sense only for k-rectifiable sets, namely sets given by a countable union of k-dimensional Lipschitz manifolds.

Second, very little progress has been made in the study of functional spaces over varifolds or currents, so that no consolidated technique allows at present to treat problems involving an equation on these weak objects.

The pioneering work which opened a possible way to manage the above difficulties was the paper [10], where Bouchitté, Buttazzo and Seppecher introduced a new notion of tangent space to a measure, making sense for *any* positive Borel measure on $\mathbb{R}^n$; consequently they also introduced a suitable notion of Sobolev functions with respect to a measure. These tools were applied in order to model the elastic energy of low-dimensional structures, such as beams or membranes. In fact, every subset S of $\mathbb{R}^n$ with Hausdorff dimension $\alpha \in \mathbb{R}$ may be identified with a positive measure μ, given by the overlying measure $\mathcal{H}^\alpha$, possibly weighted by a positive density θ; more in general, a multijunction made by the union of sets S_i with different dimensions α_i may be described through a positive measure μ of the kind $\sum_i \theta_i \mathcal{H}^{\alpha_i} \llcorner S_i$.

From [10], several successive works have originated, where the new tools are studied and developed, and the idea to use measures for describing weak geometric structures in $\mathbb{R}^n$ is applied to different kinds of variational problems. We indicate shape optimization and homogenization as the areas where the measure approach has allowed to find out the most interesting results; also, optimal control problems with manifolds as controls can be considered in this framework.

The present paper collects, in the form of a self-contained survey, the main results obtained until now by the measure method. Besides proofs, many details are often avoided, in order to offer a not too technical presentation. We also stress that the significant part relative to shape optimization and mass transport is omitted herein; we refer for that to [11], [6] and also to the paper by Buttazzo, Oudet and Stepanov included in this volume.

The outline of the paper is the following. In Section 2 we summarize the basic notions of geometric kind, such as the tangent space and the mean curvature of a measure μ. In Section 3 we treat functional spaces: we introduce Sobolev functions of type H and W, and functions with bounded variation, all with respect to μ. Applications to variational problems are described in Section 4: Subsection 4.1 deals with the semicontinuity and relaxation of integral energies; Subsection 4.2 deals with the homogenization of periodic low-dimensional structures; Subsection 4.3 deals with minimum and optimal control problems involving manifolds. Some of the results contained in Subsections 4.1 and 4.2 are new, and related papers are still to appear.

2. Differential geometry of μ

Throughout the paper, unless otherwise specified, μ is a positive Borel measure on $\mathbb{R}^n$, and integrals are extended over $\mathbb{R}^n$.

2.1. Tangent space

Any measure μ as above admits a tangent space at μ-a.e. point $x \in \mathbb{R}^n$. It is a linear subspace of $\mathbb{R}^n$, which will be denoted by $T_\mu(x)$. Before giving the definition of $T_\mu(x)$, let us briefly explain where it comes from. It is based on the following linear operators approach.

For a fixed exponent $p \in [1, +\infty]$, set for brevity $L^p_\mu := L^p_\mu(\mathbb{R}^n)$, and $(L^p_\mu)^n := L^p_\mu(\mathbb{R}^n; \mathbb{R}^n)$; we use the common notation $\|\cdot\|_{p,\mu}$ for the respective norms. Consider then the gradient operator $G : L_\mu^p \to (L_\mu^p)^n$; it is densily defined by setting $Gu := \nabla u$ on the space $\mathcal{D}$ of compactly supported smooth functions. Therefore, one can consider the adjoint operator $G^* : (L_\mu^{p'})^n \to L_\mu^{p'}$ (being p' the conjugate exponent such that $\frac{1}{p} + \frac{1}{p'} = 1$). It is easily checked that the domain of G^* is the space of vector fields

$$X_\mu^{p'} := \left\{ \Phi \in (L_\mu^{p'})^n \; : \; \mathrm{div}(\Phi\mu) \in L_\mu^{p'} \right\} ,$$

where $\mathrm{div}(\Phi\mu)$ is intended in the sense of distributions, i.e., $\langle \mathrm{div}(\Phi\mu), \psi \rangle := -\int_{\mathbb{R}^n} \Phi \cdot \nabla\psi \, d\mu$ for all $\psi \in \mathcal{D}$. For every $\Phi \in X_\mu^{p'}$, $G^*\Phi$ equals $-\mathrm{div}(\Phi\mu)$.

Note that, when μ is the Hausdorff measure $\mathcal{H}^k$ over a smooth k-dimensional manifold S, taking a nonzero test function ψ which vanishes on S, the regularity condition on $\mathrm{div}(\Phi\mu)$ entails that every field $\Phi \in X_\mu^{p'}$ must be *tangent* to S. Whence the idea to consider, at a given point x, the envelope of all vectors $\Phi(x)$, for Φ running over the class of tangent fields $X_\mu^{p'}$. This can be made rigorous using the concept of μ-essential union: by definition, T_μ is the smallest μ-measurable and closed-valued multifunction, from $\mathbb{R}^n$ into the class of all subsets of $\mathbb{R}^n$, such that the inclusion $\{\Phi(x) : \Phi \in X_\mu^{p'}\} \subseteq T_\mu(x)$ holds for μ-a.e. x. For the existence of such minimal multifunction, called the μ-*essential union of the family* $X_\mu^{p'}$, see [31], Proposition 1.14.

We list below some properties of T_μ. More detailed results and their proofs can be found in [10], [24].

Examples. If $\mu = \mathcal{H}^k \llcorner S$, with S a k-dimensional Lipschitz manifold in $\mathbb{R}^n$ ($k \leq n$), then T_μ coincides μ-a.e. with the usual tangent bundle T_S given by Differential Geometry. If μ is as above but S is simply a k-rectifiable set, then T_μ is contained μ-a.e. into the approximate tangent space to S defined by blow-up (see for instance [30]). Such inclusion may be strict: for instance, if $n = 1$ and μ is the measure $\mathcal{H}^1$ over a Cantor-type set C, the approximate tangent space to C is 1-dimensional, whereas it can be proved that $T_\mu = \{0\}$. Actually, T_μ is reduced to zero for sets having suitable self-similarity properties (*e.g.* the Koch curve in $\mathbb{R}^2$), and also in the case of Dirac masses. Note that, for all examples mentioned above, T_μ is independent of the initial choice of the exponent p.

Locality properties. The multifunction T_μ is local on open subsets of $\mathbb{R}^n$, that is, $T_\mu = T_\nu$ μ-a.e. on E if $\mu \llcorner E = \nu \llcorner E$ and $E \subset \mathbb{R}^n$ is open. The same property does not hold if E is any Borel set (for $n = 1$, take E equal to a Cantor set C as above, $\mu = \mathcal{H}^1 \llcorner C$ and $\nu = \mathcal{H}^1$).

Dimension properties. The dimension of $T_\mu(x)$ as a linear subspace of $\mathbb{R}^n$ may be a nonconstant function of x, for instance when μ represents a multijunction, namely it is the sum of Hausdorff measures of different dimensions. In general, we have the inequality $\dim(T_\mu) \leq \mathcal{H}-\dim(E)$, μ-a.e. on every Borel subset E of $\mathbb{R}^n$ (here $\mathcal{H}-\dim(E)$ denotes the Hausdorff dimension of E).

Rectifiability properties. As mentioned in the introduction, a set is k-rectifiable if and only if it admits a k-dimensional approximate tangent space (see [30], Theorem 11.8). Using the notion of tangent space to a measure, it is possible to prove a rectifiability criterion for sets of varying Hausdorff dimension. It states that, for a given measure μ and for every integer $k \leq n$, the set E_k of points where T_μ has dimension k is k-rectifiable provided $\mu \llcorner E_k$ satisfies the doubling property and has a finite k-dimensional lower density. The statement in detail can be found in Theorem 4.5 of [24], whose proof is based on the relation between T_μ and the notion of *tangent measures* to μ introduced by Preiss in [29].

2.2. Mean curvature

As in case of the tangent bundle, the notion of mean curvature vector $H(\mu)$ can be given for *any* positive measure μ. Denote by $\mathcal{D}^n$ the space of smooth vector fields compactly supported on $\mathbb{R}^n$, and by P_μ the orthogonal projector from $\mathbb{R}^n$ onto T_μ; for every $X \in \mathcal{D}^n$, $X = (X^1, \ldots, X^n)$, set $\operatorname{div}_\mu X := \sum_{i=1}^n (P_\mu \nabla X^i)_i$. By definition, $H(\mu)$ is the vector distribution $\operatorname{div}(P_\mu \mu)$, which acts as

$$\langle H(\mu), X \rangle := -\int \operatorname{div}_\mu X \, d\mu \,, \qquad \forall X \in \mathcal{D}^n \,.$$

Measures μ for which the distribution $H(\mu)$ is in turn a (vector) measure will be called *measures with bounded curvature*. The class of measures with bounded curvature will be denoted by $\mathcal{M}_{BC}$. For $\mu \in \mathcal{M}_{BC}$, the Radon-Nikodym theorem yields the decomposition

$$H(\mu) = h(\mu)\mu + \partial\mu \,.$$

Here $h(\mu) \in (L^1_\mu)^n$ is the density of $H(\mu)$ with respect to μ, and the singular part $\partial\mu$ is a measure orthogonal to μ, that we call the *boundary* of μ. This notion of boundary is useful to formulate Dirichlet kind conditions in variational problems, see Subsection 4.3.

We give below some examples and remarks about measures in $\mathcal{M}_{BC}$, referring to [7] for more details.

Examples. If $\mu = \mathcal{H}^k \llcorner S$, with S a k-dimensional manifold of class C^2 in $\mathbb{R}^n$, the tangential divergence theorem yields $H(\mu) = h(S)\mathcal{H}^k \llcorner S + \eta(S)\mathcal{H}^{k-1} \llcorner \partial S$, where $h(S)$ is the classical, pointwise defined, mean curvature vector, and $\eta(S)$ is the unit conormal vector to the boundary of S (*i.e.*, the normal vector to ∂S lying in the tangent bundle to S). If a measure μ of the kind above is multiplied by a smooth positive density θ, we have $H(\theta\mu) = \theta H(\mu) + P_\mu(\nabla\theta)\mu$. If μ is still of the kind above, but S is only piecewise C^2, the corresponding mean curvature can be found by applying the divergence theorem separately on each smooth component of S and then summing. This may give contributions to the boundary of μ, even if S has no boundary. For instance, if $k = 1$ and S is the boundary of a triangle, then S has no boundary, whereas the mean curvature of μ is purely singular, and it is given by three vector Dirac masses concentrated at the vertices. Finally, if μ is still of the kind $\theta\mathcal{H}^k \llcorner S$, but S is only k-rectifiable, the reader who is familiar with the theory of rectifiable varifolds will see that the definition of $H(\mu)$ is similar to the notion of first variation δV for the varifold $V = (S, \theta)$ associated to μ ([30], Section 39). It can be proved that $H(\mu)$ coincides with δV under the assumption that δV is a vector measure (this is in general not equivalent to require that $\mu \in \mathcal{M}_{BC}$, unless T_μ coincides with the approximate tangent space to S, see [7], Example 2.8).

Remarks. The notion of mean curvature for arbitrary measures reveals useful in connection with the new features mentioned in the introduction: functional spaces and varying dimension. Indeed, a distributional notion of Sobolev functions with respect to μ can be given only under suitable regularity conditions on $H(\mu)$ (see Subsection 3.2): this link between Sobolev spaces and mean curvature is rather surprising and is completely hidden in the usual case of the Lebesgue measure. The regularity of $H(\mu)$ is also needed for the integral representation of relaxed energies with linear growth on the space BV_μ (see Subsection 4.1). Further, when dealing with varying measures $\{\mu_n\}$, the passage to the limit in the sequence of the dimensions $\{\dim T_{\mu_n}\}$ is strictly related to the passage to the limit in the sequence of the curvatures $\{H(\mu_n)\}$ (see Subsection 4.3); thus, the class $\mathcal{M}_{BC}$ suits the aim of studying the stability of upper or lower constraints on the dimension of geometric structures.

3. Functional spaces associated to μ

In the well-known case of the Lebesgue measure, the definition of strong and weak Sobolev spaces (of type "H" and "W" respectively), is founded upon two different concepts of gradient: the pointwise one and the distributional one; the latter allows also to consider functions with bounded variation. Let us see how this can be transposed to the case of an arbitrary measure μ.

3.1. Sobolev spaces of type H

Fix an exponent $p \in [1, +\infty]$. The definition of $H_\mu^{1,p}$ is based on the pointwise notion of tangential gradient with respect to μ. For every smooth function u, it is obtained simply by projecting the usual gradient ∇u onto the tangent space T_μ:

$$\nabla_\mu u := P_\mu(\nabla u) \qquad \forall u \in \mathcal{D} .$$

Consider now a sequence $\{u_n\} \subset \mathcal{D}$ such that $u_n \to u$ in L_μ^p and $\{\nabla_\mu u_n\}$ is bounded in $(L_\mu^p)^n$. Saying that u is a Sobolev function will make sense provided the sequence $\{\nabla_\mu u_n\}$ admits a unique limit, which will represent the μ-tangential gradient of u. In other words, we are asking whether the μ-tangential gradient is closable as a linear operator from $\mathcal{D} \subset L_\mu^p$ into $(L_\mu^p)^n$. We recall that a linear operator $A : X \to Y$ between Banach spaces, with dense domain $D(A) \subset X$, is said to be *closable* if

$$u_n \to 0 \text{ in } X , \quad Au_n \to v \text{ in } Y \quad \Rightarrow \quad v = 0 . \tag{1}$$

This means that the closure of the graph $G(A)$ of A is in turn the graph of a linear operator $\overline{A}$, characterized by the equality $\overline{G(A)} = G(\overline{A})$. In particular, the domain of $\overline{A}$ is obtained projecting $G(\overline{A}) \subset X \times Y$ onto the first factor X.

Note that, for μ equal to the Lebesgue measure $\mathcal{L}^n$, when $X = L^p_{dx}$, $Y = (L^p_{dx})^n$, $D(A) = \mathcal{D}$, and $Au = \nabla u$, the validity of (1) follows from the integration by parts formula

$$\int u_n \nabla \psi \, dx = -\int \psi \nabla u_n \, dx \qquad \forall \psi \in \mathcal{D} \,,$$

combined with the implication

$$v \in (L^p_\mu)^n \,, \quad \int \psi v \, dx = 0 \quad \forall \psi \in \mathcal{D} \Rightarrow v = 0 \text{ a.e.}$$

In case of an arbitrary measure μ, when $X = L^p_\mu$, $Y = (L^p_\mu)^n$, $D(A) = \mathcal{D}$, and $Au = \nabla_\mu u$, the closability property (1) still holds thanks to the integration by parts formula

$$\langle u_n, \operatorname{div}(\Phi \mu) \rangle = -\int \Phi \cdot \nabla_\mu u_n \, d\mu \qquad \forall \Phi \in X^{p'}_\mu \,, \tag{2}$$

combined with the implication

$$v \in (L^p_\mu)^n \,, \quad \int \Phi \cdot v \, d\mu = 0 \quad \forall \Phi \in X^{p'}_\mu \,, \quad v(x) \in T_\mu(x) \ \mu\text{-a.e.} \Rightarrow v = 0 \text{ a.e.} \tag{3}$$

The validity of both (2) and (3) derives from the construction of the tangent space T_μ as a multifunction related to the gradient operator G. With the same notation as in Subsection 2.1, (2) follows from the identity $P_\mu \Phi = \Phi$, holding for all $\Phi \in D(G^*) = X^{p'}_\mu$, while (3) is based on a commutation argument between integral and μ-essential supremum (see [15], Theorem 1). Such argument works thanks to the following *locality property* of $D(G^*)$:

$$\Phi \in D(G^*) \,, \quad \psi \in \mathcal{D} \quad \Rightarrow \quad \psi \Phi \in D(G^*) \,. \tag{4}$$

Let us stress for later use how the above implication can be verified. Note first that the locality property is stable by passage to the orthogonal complement, and recall the following identity from Functional Analysis:

$$D(G^*)^\perp = \overline{G}(0) := \{\xi \in (L^p_\mu)^n : \exists u_h \subset D(G) \text{ such that } u_h \to 0 \text{ and } Gu_h \to \xi\} \,.$$

Then, take $\xi \in \overline{G}(0)$ and $\psi \in \mathcal{D}$. If $\{u_h\} \subset \mathcal{D}$ satisfies $u_h \to 0$ and $Gu_h \to \xi$, the sequence $\{\psi u_h\}$ satisfies $\psi u_h \to 0$ and $G(\psi u_h) \to \psi \xi$. Hence $\psi \xi \in \overline{G}(0)$ and (4) is proved.

Summing up, thanks to (2) and (3), we have that A is closable (the complete proof can be found in [10]). We define $H^{1,p}_\mu := D(\overline{A})$. By construction, $H^{1,p}_\mu$ is a Banach space. In analogy to the Lebesgue case, $H^{1,p}_\mu$ turns out to be reflexive for $p \in (1, +\infty)$ and separable for $p \in [1, +\infty)$, and suitable characterizations by duality and by relaxation hold. In fact, for a function $u \in L^p_\mu$ the following facts are equivalent when $p > 1$:

$$u \in H^{1,p}_\mu \,; \tag{5}$$

$$\exists C > 0 : |\langle u, \operatorname{div}(P_\mu \Phi \mu) \rangle| \le C \|\Phi\|_{p',\mu} \ \forall \Phi \in (L^{p'}_\mu)^n \text{ with } P_\mu \Phi \in X^{p'}_\mu \,; \tag{6}$$

$$\exists \{u_h\} \subset \mathcal{D} : u_h \to u \text{ in } L^p_\mu \text{ with } \nabla u_h \text{ bounded in } (L^p_\mu)^n \,. \tag{7}$$

Variants of the space $H_\mu^{1,p}$, useful to treat different variational problems, can be easily obtained by suitably changing the domain of the linear operator $Au = \nabla_\mu u$. We have

(i) for $X = L_\mu^p(\Omega)$, $Y = (L_\mu^p(\Omega))^n$, $D(A) = \mathcal{D}$, then $D(\overline{A})$ is the space $H_\mu^{1,p}(\Omega)$ of Sobolev functions on an open subset Ω of $\mathbb{R}^n$;

(ii) for $X = L_\mu^p(\Omega)$, $Y = (L_\mu^p(\Omega))^n$, $D(A) = \mathcal{D}(\Omega)$, then $D(\overline{A})$ is the space $H_{\mu,0}^{1,p}(\Omega)$ of Sobolev functions vanishing on $\partial\Omega$;

(iii) when μ is a periodic measure, for $X = L_{\mu,\mathrm{loc}}^p$, $Y = (L_{\mu,\mathrm{loc}}^p)^n$, $D(A) - C_\sharp^\infty$ (the space of periodic smooth functions with the same unit cell as μ), then $D(\overline{A})$ is the space $H_{\mu,\sharp}^{1,p}$ of periodic Sobolev functions.

Case of elasticity. All the above definitions can be extended in the natural way to the case of vector-valued functions. For applications to the elasticity setting, it is also useful to consider the notion of μ-*tangential strain*. This can be done simply replacing the gradient operator G by its symmetric part. More precisely, define $Eu = \frac{\nabla u + \nabla^T u}{2}$ as a linear operator from $\mathcal{D}^n \subset (L_\mu^p)^n$ into the space $(L_\mu^p)_{\mathrm{sym}}^{n^2}$ of L_μ^p-functions with values in the symmetric $n \times n$ real matrices $\mathbb{R}_{\mathrm{sym}}^{n^2}$. Set $L_\mu(x) = \mu-\mathrm{ess}\bigcup\{\Phi(x) : \Phi \in D(E^*)\}$, and denote by Π_μ the orthogonal projector from $\mathbb{R}_{\mathrm{sym}}^{n^2}$ onto L_μ. For measures supported on smooth manifolds, L_μ turns out to be linked with T_μ by the relation $\Pi_\mu \sigma = P_\mu \sigma P_\mu$, holding for all $\sigma \in \mathbb{R}_{\mathrm{sym}}^{n^2}$. Letting now $Au := \Pi_\mu(eu)$, since $D(E^*)$ enjoys the locality property (4), we can prove that A is closable, and:

(iv) for $X = (L_\mu^p(\Omega))^n$, $Y = (L_\mu^p(\Omega))_{\mathrm{sym}}^{n^2}$, $D(A) = \mathcal{D}^n$, then $D(\overline{A})$ is the space $\mathcal{D}_\mu^{1,p}(\Omega)$ of admissible displacements on an open subset Ω of $\mathbb{R}^n$;

(v) when μ is a periodic measure, for $X = (L_{\mu,\mathrm{loc}}^p)^n$, $Y = (L_{\mu,\mathrm{loc}}^p)_{\mathrm{sym}}^{n^2}$, $D(A) = (C_\sharp^\infty)^n$, then $D(\overline{A})$ is the space $\mathcal{D}_{\mu;\sharp}^{1,p}$ of periodic admissible displacements.

Case of the Hessian operator. One can ask at this point whether the same completion process can be employed to define higher order Sobolev spaces. The natural starting point would be in this case the hessian matrix operator $Hu = \nabla^2 u$ from $\mathcal{D} \subset L_\mu^p$ into $(L_\mu^p)_{\mathrm{sym}}^{n^2}$. Such operator H presents a crucial difference with respect to the first order gradient G: the domain of the adjoint $D(H^*)$ (or equivalently $\overline{H}(0)$) does not enjoy the locality property (4). Indeed, if $\xi \in \overline{H}(0)$ and $\psi \in \mathcal{D}$, let $\{u_h\} \subset \mathcal{D}$ satisfy $u_h \to 0$ and $\nabla^2 u_h \to \xi$. Taking the sequence $\{\psi u_h\}$ is not sufficient to ensure that the product $\psi\xi$ belongs to $\overline{H}(0)$. This is due to the first order terms in the expansion

$$\nabla^2(\psi u_h) = \psi \nabla^2 u_h + \nabla\psi \otimes \nabla u_h + \nabla u_h \otimes \nabla\psi + u_h \nabla^2 \psi .$$

In order to gain a control on first order derivatives by means of the hessian, different strategies are needed for the tangential and the orthogonal parts of the gradient.

The tangential part can be controlled in terms of second derivatives by imposing that μ satisfies a p-Poincaré inequality of the kind:

$$\exists\, C > 0 : \int |u - \overline{u}|^p \, d\mu \le C \int |\nabla_\mu u|^p \, d\mu \quad \forall u \in \mathcal{D}, \text{ where } \overline{u} := \frac{1}{|\mu|} \int u \, d\mu \ . \quad (8)$$

The orthogonal part $b := P_\mu^\perp \nabla u$ (where $P_\mu^\perp$ denotes the orthogonal projector on $T_\mu^\perp$) turns out to play a significant role, in some point similar to that of *Cosserat vector field* for bending theory in elasticity. By taking into account this Cosserat vector field, we extend the domain of the Hessian operator considered on the product space $L_\mu^p \times L_\mu^p(\mathbb{R}^n; T_\mu^\perp)$ by setting:

$$D(H) = \{(u, b) : u \in \mathcal{D}, \, b = \nabla u - \nabla_\mu u\}, \quad H(u, b) = \nabla^2 u \ .$$

With the above definition of H, and under the Poincaré assumption (8) on μ, one can prove that $D(H^*)$ satisfies the locality property. Then we set $M_\mu(x) = \mu\text{-ess}\bigcup\{\Phi(x) : \Phi \in D(H^*)\}$, and we denote by Q_μ the orthogonal projector from $\mathbb{R}_{\text{sym}}^{n^2}$ onto M_μ. For measures μ concentrated on smooth manifolds, M_μ turns out to be linked with T_μ by the equality $Q_\mu \sigma = P_\mu \sigma P_\mu + P_\mu \sigma P_\mu^\perp + P_\mu^\perp \sigma P_\mu$, holding for all $\sigma \in \mathbb{R}_{\text{sym}}^{n^2}$, where $P_\mu^\perp$ is the orthogonal projector on $T_\mu^\perp$. For $(u, b) \in D(H)$, define now $A(u, b) := Q_\mu(\nabla^2 u)$. Thanks to the locality of $D(H^*)$, the operator A results closable, thus providing a μ-intrinsic object containing second order derivatives. The structure of the operator $\overline{A}$ is rather complicated, as it involves a curvature tensor related to μ. In particular, it is not immediate to characterize the domain of $\overline{A}$ and to deduce the natural definition of second order Sobolev spaces associated to μ. With some lack of precision, we may say that pairs (u, b) in $D(\overline{A})$ must satisfy the condition $\nabla_\mu u + b \in (H_\mu^{1,p})^n$. We limit here to describe the operator $\overline{A}$ in the simple example when μ is the measure $\mathcal{H}^1$ over a smooth closed curve S in $\mathbb{R}^2$. In this case, we have $D(\overline{A}) = H^2(S) \times (H^1(S, T_S^\perp))^2$. Moreover, denoting by u', u'' first and second derivation with respect to the arc parameter, and by K the scalar mean curvature of S, in the moving frame given by the tangent and the normal vectors τ and ν to S, the matrix $\overline{A}(u, b\nu)$ takes the form

$$\overline{A}(u, b\nu) = (u'' - Kb)(\tau \otimes \tau) + (b' + Ku')(\tau \otimes \nu + \nu \otimes \tau)$$

(here $b \in H^1(S)$ denotes a *scalar* function). We refer to [14] for the general definitions of μ-tangential hessian and related Sobolev spaces $H_\mu^{2,p}$.

3.2. Sobolev spaces of type W

For a fixed exponent $p \in [1, +\infty]$, the definition of $W_\mu^{1,p}$ is based on the distributional notion of tangential gradient with respect to μ. It can be introduced for those measures $\mu \in \mathcal{M}_{BC}$ whose mean curvature satisfies the regularity conditions

$$h(\mu) \in (L_\mu^\infty)^n, \quad \mu(\text{spt}\partial\mu) = 0, \quad (9)$$

where spt$\partial\mu$ denotes the support of the vector measure $\partial\mu$. Under the assumption (9) on μ, the μ-tangential gradient of a function $u \in L_\mu^p$ in the sense of distributions can be defined as

$$D_\mu u := \text{div}(uP_\mu\mu) \lfloor (\mathbb{R}^n \setminus \text{spt}\partial\mu) - uh(\mu)\mu .$$

Note that, when u is smooth, we have $\text{div}(uP_\mu\mu) = uH(\mu) + (\nabla_\mu u)\mu$. Therefore, the distributional equality $D_\mu u = (\nabla_\mu u)\mu$ holds for $u \in \mathcal{D}$, and it can be extended by density to all $u \in H_\mu^{1,p}$. Note also that, when $\mu = \mathcal{L}^n$, $D_\mu u$ reduces to the usual notion of distributional gradient of u. We set

$$W_\mu^{1,p} := \{u \in L_\mu^p : D_\mu u \in (L_\mu^p)^n\} .$$

It is not difficult to prove that $W_\mu^{1,p}$ is a Banach space, endowed with the norm $\|u\|_{p,\mu} + \|\nabla_\mu u\|_{p,\mu}$. To that aim, both conditions in (9) intervene: the former implies that the request $D_\mu u \in (L_\mu^p)^n$ is equivalent to $\text{div}(uP_\mu\mu) \lfloor (\mathbb{R}^n \setminus \text{spt}\partial\mu) \in (L_\mu^p)^n$; the latter entails that, for $u \in W_\mu^{1,p}$, the equality $\text{div}(uP_\mu\mu) \lfloor (\mathbb{R}^n \setminus \text{spt}\partial\mu) = \text{div}(uP_\mu\mu)$ holds in $(L_\mu^p)^n$.

Considering distributions on Ω instead of $\mathbb{R}^n$, one can similarly define the space $W_\mu^{1,p}(\Omega)$. Also, a Dirichlet condition can be imposed on $\partial\Omega$ by setting $W_{\mu,0}^{1,p}(\Omega) = \{u \in L_\mu^p(\Omega) : \tilde{u} \in W_\mu^{1,p}\}$, where $\tilde{u}$ denotes the extension by zero of u outside Ω.

We remark that the equality $D_\mu u = (\nabla_\mu u)\mu$ for $u \in H_\mu^{1,p}$, yields that $H_\mu^{1,p}$ is contained as a (closed) subspace into $W_\mu^{1,p}$, and similarly $H_\mu^{1,p}(\Omega) \subseteq W_\mu^{1,p}(\Omega)$. The strict inclusion may hold, due to the possible lack of regularity either of Ω or of μ. For instance, let $n = 2$, $\mu = \mathcal{L}^2$, and $\Omega = (0,1)^2 \cup (-1,0)^2$. In this case, $H_\mu^{1,p}(\Omega)$ and $W_\mu^{1,p}(\Omega)$ coincide respectively with the classical spaces $H^{1,p}(\Omega)$ and $W^{1,p}(\Omega)$, and it is well known that the phenomenon "$H \neq W$" occurs for all $p > 2$, due to the lack of regularity of $\partial\Omega$. On the other hand, consider the case when $\Omega = \mathbb{R}^n$ (for arbitrary $n > 1$), and μ is the measure $\mathcal{H}^1$ over a smooth closed curve having a double point P, i.e., a point of auto-intersection. In this case, functions in $H_\mu^{1,p}(\Omega)$ must be continuous at P, whereas the space $W_\mu^{1,p}(\Omega)$ contains any u which, seen as a function of the arc parameter on S, belongs to the classical unidimensional Sobolev space. Hence in this case the phenomenon "$H \neq W$" is due to the change of multiplicity occurring in the support of μ at the point P. It is a puzzling open problem to establish sharp sufficient conditions on both μ and Ω in order to have the equality $H_\mu^{1,p}(\Omega) - W_\mu^{1,p}(\Omega)$.

3.3. Functions with bounded variation

When $\mu = \mathcal{L}^n$, it is known that, for $p = 1$, the equivalence between conditions (5)–(6)–(7) fails. Actually, one still has (5)$\Rightarrow$(6)$\Leftrightarrow$(7), but the converse implication (5)$\Leftarrow$(6) becomes false. The equivalent conditions (6) or (7) mean that u belongs to the space BV of functions with bounded variation, which contains properly $H^{1,1}$. An extensive study of the space BV, which has been exploited in the last years in connection with free boundary problems, can be found in the recent monograph [3]. In case of a general measure μ, one can prove that the co-implication (6)$\Leftrightarrow$(7) still holds when $p = 1$ (and $p' = \infty$). Then, we call BV_μ the space of functions

$u \in L^1_\mu$ which satisfy one of the following equivalent conditions:

$$\exists C > 0 \; : \; |\langle u, \mathrm{div}(P_\mu \Phi \mu)\rangle| \leq C \|\Phi\|_{\infty,\mu} \; \forall \Phi \in (L^\infty_\mu)^n \text{ with } P_\mu \Phi \in X^\infty_\mu \; ; \quad (10)$$

$$\exists \{u_h\} \subset \mathcal{D} \; : \; u_h \to u \text{ in } L^1_\mu \text{ with } \nabla u_h \text{ bounded in } (L^1_\mu)^n . \quad (11)$$

If we define the μ-total variation of any function $u \in L^1_\mu$ as

$$|D_\mu u| := \sup\{|\langle u, \mathrm{div}(\Phi \mu)\rangle| \; : \; \Phi \in X^\infty_\mu \; , \; \|\Phi\|_{\infty,\mu} \leq 1\} \; , \quad (12)$$

by (10) we may identify BV_μ to the class of functions $u \in L^1_\mu$ such that $|D_\mu u| < +\infty$. This allows to prove that BV_μ reduces to the classical space BV when $\mu = \mathcal{L}^n$. However, for general μ, the problem of determining a vector measure $D_\mu u$ for which the scalar positive quantity $|D_\mu u|$ represents the total variation turns out to be quite delicate. Indeed, to that aim one has to consider the linear operator $Au := (\nabla_\mu u)\mu$ defined on $D(A) := \mathcal{D} \subset L^1_\mu$, with values in the space of vector Radon measures, endowed with the weak star topology. Such operator results closable if and only if the class

$$\mathcal{V}_\mu := \{\Phi \in C^n_0 \; : \; P_\mu \circ \Phi \in X^\infty_\mu\} \quad (13)$$

is dense in the space C^n_0 of continuous vector fields vanishing at infinity on $\mathbb{R}^n$, endowed with the uniform norm. Under this density condition, the operator $\overline{A}$ is well defined, and its domain turns out to contain BV_μ. Then we can set $D_\mu u := \overline{A}u$ for all $u \in BV_\mu$, and the total variation of the measure $D_\mu u$ is given by the right-hand side of (12). In particular, for $u \in H^{1,1}_\mu$, we have $D_\mu u = (\nabla_\mu u)\mu$.

The natural question is now the following: for what measures μ the density condition on $\mathcal{V}_\mu$, and hence the closability of A, is fulfilled? A sufficient condition, satisfied for instance when $\mu = \theta \mathcal{H}^k \llcorner S$ with θ and S smooth, is that the mean curvature of μ belongs to $(L^\infty_\mu)^n$. When $H(\mu) \notin (L^\infty_\mu)^n$, the closability of A may fail. This happens even for very simple measures, for instance when $n = 2$ and

$$\mu = \mathcal{H}^1 \llcorner S, \quad \text{with } S = S_+ \cup S_- \; , \quad S_\pm = \{(\pm t, t) : t \in [0,1]\} . \quad (14)$$

As a consequence, we have surprising formulae for the relaxation of anisotropic energies in $d\mu$ with linear growth, see Subsection 4.1.

Among other properties of BV_μ functions, we wish to put in evidence the validity of a very general coarea formula, which extends the classical one for BV functions (see [30], Section 5.5). For a μ-measurable set $E \subset \mathbb{R}^n$, we denote by χ_E its indicatrix function, we set $P_\mu(E) := D_\mu \chi_E$, and we say that E has a finite perimeter with respect to μ if $P_\mu(E) < +\infty$. Then, for all $u \in L^1_\mu$ and $\mathcal{L}^1$-a.e. $t \in \mathbb{R}$, the sublevel $\{x \in \mathbb{R}^n : u(x) < t\}$ has a finite perimeter with respect to μ, and we have the following coarea formula:

$$|D_\mu u| = \int_{\mathbb{R}} P_\mu\big(\{x \in \mathbb{R}^n : u(x) < t\}\big) \, dt \qquad \forall u \in L^1_\mu . \quad (15)$$

The above equality must be intended in the sense $+\infty = +\infty$ if $u \notin BV_\mu$, and contains implicitly the statement that the map $t \mapsto P_\mu\big(\{x \in \mathbb{R}^n : u(x) > t\}\big)$ is $\mathcal{L}^1$-measurable. Differently from the classical case $\mu = \mathcal{L}^n$, the proof is not based on approximation by smooth functions, but it is suggested by the distributional definition (12), and relies on a commutation argument between supremum and integral (see [5], Theorem 4.1).

4. Applications to variational problems

The mathematical tools introduced in the previous sections have allowed a new approach to different kinds of variational problems, where measures may intervene either as data or as unknown. Problems involving a fixed given measure, which describes the geometry of the structure under consideration, are the modelling of the elastic energy of thin structures, and the homogenization of scalar or vector problems on periodic sets of low dimension in $\mathbb{R}^n$. In spite, measures play the role of unknown when dealing with minimum or optimal control problems suggested by Differential Geometry or Mechanics, where one has to find the best manifold according to some criterion. In the remaining of the paper, we describe the main results concerning these different topics.

4.1. Energies with respect to μ: semicontinuity and relaxation

The modelling of the elastic energy of thin structures is a classical problem in Continuum Mechanics, for which justifications from a variational point of view have been searched in recent years. The most commonly used approach is the so-called passage $3D \to 2D$ or $3D \to 1D$ (see [27] and [28]). For instance, in the case of a $2D$ set in $\mathbb{R}^3$, one considers a thickness positive parameter δ, and tries to obtain the membrane energy of the structure of thickness $\delta = 0$ by means of a Γ-convergence technique, namely starting from the $3D$ energies of the approximating sets (suitably rescaled), and letting δ tend to zero. This fattening approach creates technical problems when the structure under consideration is very irregular, or contains parts of different dimensions. Therefore, it seems useful to adopt a more intrinsic method, which consists in identifying a thin structure S with a positive measure μ, given by the overlying Hausdorff measure (or sum of Hausdorff measures in case S is a multijunction). Then, if S is governed as a membrane by a second order equation, we consider an integral functional F in $d\mu$ which depends on the first order gradient, and thus is initially defined only for smooth functions. A suitable relaxation of F allows to obtain a lower semicontinuous functional $\overline{F}$, which represents the energy of the structure S associated with μ. More precisely, we have the following theorem, proved in [10]. We recall that the relaxed functional of a given functional F with respect to the L^p_μ-topology is by definition the greatest L^p_μ-lower semicontinuous functional $\overline{F}$ satisfying the inequality $\overline{F} \leq F$ [16].

Theorem 4.1. *Let*

$$F(u) = \int f(x, \nabla u)\, d\mu\,, \qquad u \in \mathcal{D}\,, \tag{16}$$

where the integrand $f = f(x, z)$ is assumed to be μ-measurable in x and convex with growth conditions of order $p > 1$ in z $(c|z|^p \leq f(x, z) \leq C(1 + |z|^p))$. The *relaxed functional of F with respect to the L^p_μ-topology is given by*

$$\overline{F}(u) = \int f_\mu(u, \nabla_\mu u) \, d\mu \, , \qquad u \in H^{1,p}_\mu \, ,$$

where the integrand f_μ is related to f by the formula

$$f_\mu(x, z) = \inf \left\{ f(x, z + \xi) \ : \ \xi \in T_\mu(x)^\perp \right\} \, . \tag{17}$$

We stress that in the above statement the functionals F and $\overline{F}$ are implicitly extended to $+\infty$ out of their domains. In particular, as it is well known for $\mu = \mathcal{L}^n$, the relaxation of the functional $\int |\nabla u|^p \, d\mu$ defined on $\mathcal{D}$ gives the energy $\int |\nabla_\mu u|^p \, d\mu$, with finiteness domain $H^{1,p}_\mu$. For general f as in the assumptions, when μ is concentrated on a smooth set, the expression of f_μ given by (17) coincides with the energy density found by approximation and reduction of dimension. Thus, since no regularity assumption is needed on the support of μ, the fattening approach is encompassed by the measure method.

Theorem 4.1 suggests also the study of lower semicontinuity and relaxation properties, in analogy to the known case $\mu = \mathcal{L}^n$. For instance, one can ask what kind of conditions on the integrand j are related to the weak lower semicontinuity on the space of vector-valued Sobolev functions $H^{1,p}_\mu(\mathbb{R}^n, \mathbb{R}^m)$ of an integral functional J of the kind

$$J(u) = \int j(x, \nabla_\mu u) \, d\mu \, , \qquad u \in H^{1,p}_\mu(\mathbb{R}^n, \mathbb{R}^m) \, . \tag{18}$$

When $\mu = \mathcal{L}^n$, this is a central problem in the Calculus of Variations. Results by now classical due to Tonelli and Ioffe in the scalar case $m = 1$, and to Morrey and later refinements by Dacorogna and Marcellini in the vector case $m > 1$, tell that the sharp condition for the weak lower semicontinuity of J is respectively the convexity and the quasiconvexity of j (see for instance the book [19]). For an arbitrary measure μ, this problem has been studied in [23], where a notion of quasiconvexity along the directions of T_μ has been introduced. Precisely, we say that a μ-measurable integrand j defined on $\mathbb{R}^n \times \mathbb{R}^{nm}$ is μ-*quasiconvex* if the following inequality holds for μ-a.e. $x \in \mathbb{R}^n$, for every linear mapping A from $T_\mu(x)$ into $\mathbb{R}^m$, and for all $\mathbb{R}^m$-valued smooth functions ψ compactly supported on the unit ball B of $\mathbb{R}^n$:

$$j(x, A) \le \left(\mathcal{H}^{k(x)}(B \cap T_\mu(x)) \right)^{-1} \int_{B \cap T_\mu(x)} j(x, A + \nabla_\mu \psi) \, d\mathcal{H}^{k(x)} \tag{19}$$

(here $k(x)$ is the dimension of $T_\mu(x)$ as a linear subspace of $\mathbb{R}^n$). When $\mu = \mathcal{L}^n$, (19) becomes the classical notion of quasiconvexity. In particular, in the scalar case $m = 1$, condition (19) reduces to the convexity of $j(x, \cdot)$ on $T_\mu(x)$. The relation between (19) and weak lower semicontinuity is expressed by the following

Theorem 4.2. *Let $j = j(x, z)$ be μ-measurable in x, Lipschitz in z, and satisfying*

$$- c(1 + |z|^q) \le j(x, z) \le C(1 + |z|^p) \text{ for some } p \in (1, +\infty) \, , \ q \in [1, p) \, ;$$
$$|j(x, z) - j(y, z)| \le \eta(|y - x|)(1 + |z|^p) \text{ for some function } \eta \text{ continuous at } 0 \, .$$

If the functional J defined by (18) is weakly lower semicontinuous on $H^{1,p}_\mu(\mathbb{R}^n,\mathbb{R}^m)$, then the integrand j is μ-quasiconvex. The converse implication holds under the assumption that μ is a measure of the kind $\theta \mathcal{H}^k \llcorner S$, where S is a k-rectifiable set and θ is a nonnegative function locally $\mathcal{H}^k$-integrable on S, satisfying in addition a p-Poincaré inequality of the kind (8) on every ball $B_\rho(x)$ for μ-a.e. x and for all $\rho > 0$.

Other problems related to Theorem 4.1 are the relaxation of integral energies with linear growth (case $p = 1$), or depending on higher order derivatives (case of the hessian).

In the case $p - 1$, one has to deal with the space BV_μ, and the problem of giving an integral representation for the relaxed functional turns out to be quite delicate. Indeed, for $p > 1$, the proof of Theorem 4.1 relies on the following lemma, taking $X = L^p_\mu$, $Y = (L^p_\mu)^n$, $D(T) = \mathcal{D}$, and $Tu = \nabla_\mu u$.

Lemma 4.3. *Let $T : D(T) \subseteq X \to Y$ be a linear operator between Banach spaces, with $D(T)$ dense in X, and Y reflexive. Let $\Psi : Y \to (-\infty, +\infty]$ be a convex, coercive, and lower semicontinuous function with $\inf \Psi > -\infty$. For $u \in D(T)$, set $F(u) := \Psi(Tu)$. Then the relaxed functional of F is given by $\overline{F}(u) = \inf\{\Psi(v) : (u, v) \in \overline{G(T)}\}$.*

For $p = 1$, due to the lack of reflexivity, the above lemma cannot be applied. In fact the relaxation becomes more complicated, and suitable assumptions on both the measure and the density integrand are needed to get an integral representation. If F is defined by (16), where the integrand f satisfies the same hypotheses as in Theorem 4.1 with $p = 1$, we have the following result for the relaxed functional $\overline{F}$ with respect to the L^1_μ-topology.

Theorem 4.4. *Under the above assumptions, the finiteness domain of $\overline{F}$ is the space BV_μ. If the integrand f is continuous, for all $u \in H^{1,1}_\mu$ we have $\overline{F}(u) = \int f_\mu(x, \nabla_\mu u)\, d\mu$, with f_μ given by (17). Suppose in addition that the space V_μ defined by (13) is dense in C^n_0, and that μ satisfies the regularity condition*

$$|D_\mu u| = \sup\left\{\langle u, \operatorname{div}(P_\mu \Phi \mu)\rangle : \Phi \in V_\mu , \|\Phi\|_{\infty,\mu} \le 1\right\} \quad \forall u \in L^1_\mu .$$

Then the measure $D_\mu u$ is well defined (cf. Subsection 3.3), and we have the integral representation $\overline{F}(u) = \int f_\mu(x, D_\mu u)$ for all $u \in BV_\mu$.

We stress that the assumptions on μ in the above statement are satisfied when $\mu = \theta \mathcal{H}^k \llcorner S$, with θ and S smooth, but they mail fail even for very simple measures. For instance, if μ is the measure on $\mathbb{R}^2$ defined by (14), the integral representation on BV_μ does not hold, and it is replaced by a striking relaxation formula. Taking $f = f(z)$ independent of x, and setting $\nu_\pm = (1/\sqrt{2}, \pm\sqrt{2})$, for u equal to the indicatrix function of S_+ we have (see [5]):

$$\overline{F}(u) = \min\left\{f_\mu(\nu_+), f_\mu(\nu_-)\right\} .$$

In the case of second order energies, we are led to deal with the relaxation of the integral functional F defined by (16), when the first order gradient ∇u is replaced by the hessian matrix $\nabla^2 u$. This problem is related to the modelling of structures such as plates or rods, which are governed by a fourth order equation, and are usually treated by the method of asymptotic expansions (see [17] and references therein). For $p > 1$, the relaxation can be performed using Lemma 4.3, with $X = L_\mu^p \times L_\mu^p(\mathbb{R}^n; T_\mu^\perp)$, $Y = (L_\mu^p)_{\text{sym}}^{n^2}$, $D(T) = \{(u,b) : u \in \mathcal{D}, b = \nabla u - \nabla_\mu u\}$, $T(u,b) = \nabla^2 u$. Due to the presence of the Cosserat vector field b, the resulting relaxed energy turns out to be *non-local*, hence we have a result completely different from the first order case. As made in Subsection 3.1, we restrict attention to the simple case when μ is the measure $\mathcal{H}^1$ over a smooth closed curve S in $\mathbb{R}^2$, and we also take a simple quadratic density $f(z) = |z|^2$. We refer to the forthcoming paper [14] for the statement with general μ and f, which deserves nice trace formulae when applied to piecewise straight rods.

Theorem 4.5. *Let S be a smooth closed curve in the plane, with scalar mean curvature K. For $u \in \mathcal{D}$, let $F(u) := \int_S |\nabla^2 u|^2 \, d\mathcal{H}^1$. The relaxed functional of F with respect to the L^2-topology on S is given by*

$$\overline{F}(u) = \inf \left\{ \int_S |u'' - Kb|^2 + 2|b' + Ku'|^2 \, d\mathcal{H}^1 \ : \ b \in H^1(S) \right\}, \qquad u \in H^2(S) .$$

4.2. Homogenization of thin structures

Suppose for instance we want to homogenize a periodic array of one-dimensional bars, that is, to find the effective limit energy when the size ϵ of the periodicity cell tends to zero. Also in this case, the approach usually adopted in the literature, which is mainly due to Cioranescu and Saint-Jean Paulin [18], consists in introducing a fattening parameter $\delta > 0$, making the homogenization procedure for the associated structures of positive Lebesgue measure (namely keeping δ fixed while $\epsilon \to 0$), and finally letting $\delta \to 0$. This approach requires an extension operators' technique and several computations which are each time different according to the geometry of the structure under consideration. Again, identifying periodic thin structures with positive measures (now satisfying a periodicity condition), is fruitful to obtain more general results and some interesting applications.

Throughout this subsection, μ is a positive and periodic Borel measure on $\mathbb{R}^n$, with periodicity cell given by the unit cube Y of $\mathbb{R}^n$, and $\mu(Y) = 1$. For every $\epsilon > 0$, we denote by μ_ϵ the periodic measure obtained from μ by rescaling the size of the periodicity cell from 1 into ϵ (*i.e.* $\mu_\epsilon(B) = \epsilon^n \mu(B/\epsilon)$ for every Borel set B of $\mathbb{R}^n$). Observe that μ_ϵ converge weakly to $\mathcal{L}^n$ when $\epsilon \to 0$. Then, respectively in the scalar case (relative to the homogenization of conductors), and in the vector case (relative to the homogenization of elastic structures), we associate to μ_ϵ the following integral energies

$$F_\epsilon(u) = \int_\Omega f(\nabla u) \, d\mu_\epsilon , \qquad J_\epsilon(u) = \int_\Omega j(eu) \, d\mu_\epsilon . \qquad (20)$$

The definition domains of F_ϵ and J_ϵ are respectively the spaces of smooth compactly supported scalar functions and vector fields on the open bounded set $\Omega \subset \mathbb{R}^n$. The integrands f and j are supposed to be convex with growth conditions of order $p > 1$ from both above and below. We also associate to the sequence $\{\mu_\epsilon\}$ the following notion of convergence for functions

$$u_\epsilon \to u \quad \Longleftrightarrow \quad u_\epsilon \mu_\epsilon \rightharpoonup u \mathcal{L}^n \quad \text{weakly* as measures}, \qquad (21)$$

which arises identifying functions with μ_ϵ-absolutely continuous measures.

For $\delta > 0$, we let μ^δ be the Y-periodic measures associated to the approximating structures of thickness δ: precisely, let ρ^δ be a convolution kernel $\rho^\delta(x) := 1/\delta^n \rho(x/\delta)$, where ρ is assumed to be a smooth, positive, even function, with support compactly contained into Y, and such that $\int \rho\, dx = 1$. We set $\mu^\delta := (\rho^\delta \star \mu)\mathcal{L}^n$, being $\rho^\delta \star \mu$ the smooth function $\rho^\delta \star \mu(x) := \int \rho^\delta(x - y)\, d\mu(y)$. Observe that the measures μ^δ converge weakly to μ as $\delta \to 0$. Then, respectively in the scalar and in the vector case, we associate to μ^δ the energies F_ϵ^δ and J_ϵ^δ defined analogously to F_ϵ and J_ϵ when the ϵ-periodizations μ_ϵ of μ are replaced by the ϵ-periodizations μ_ϵ^δ of μ^δ. Similarly as above, we also associate to the sequence $\{\mu^\delta\}$ the notion of convergence for functions

$$u^\delta \to u \quad \Longleftrightarrow \quad u^\delta \mu^\delta \rightharpoonup u \mu \quad \text{weakly* as measures}. \qquad (22)$$

The behaviour of the two-parameter integrals F_ϵ^δ and J_ϵ^δ turns out to be the main difference between scalar case and vector case. Let us consider them separately.

Scalar case. We call homogenized functional of $\{F_\epsilon\}$ the energy F^{hom} obtained as the Γ-limit of F_ϵ with respect to the convergence defined by (21), that is

$$F^{\mathrm{hom}}(u) = \inf\left\{ \liminf_{\epsilon \to 0} F_\epsilon(u_\epsilon) \;:\; u_\epsilon \to u \right\} = \inf\left\{ \limsup_{\epsilon \to 0} F_\epsilon(u_\epsilon) \;:\; u_\epsilon \to u \right\}$$

(see the book [20] for an introduction to Γ-convergence). Moreover, we say that μ is *p-connected* if the following Poincaré inequality on "integer dilatations" of Y holds:

$$\exists\, C > 0 : \int_{kY} |u|^p\, d\mu \le C\, k^p \int_{kY} |\nabla_\mu u|^p\, d\mu \quad \forall k \in \mathbb{N}\,,\ \forall u \in H_{\mu,loc}^{1,p} : \int_{kY} u\, d\mu = 0,$$

where $H_{\mu,loc}^{1,p} := \{u \in L_{\mu,loc}^p \;:\; u\psi \in H_\mu^{1,p}\ \forall \psi \in \mathcal{D}\}$. Observe that, taking $k = 1$, one obtains the implication $\nabla_\mu u = 0$ μ-q.o. on $Y \Rightarrow u = 0$ μ-q.o. on Y. This motivates the choice of the terminology "p-connected". However, the p-connectedness assumption on μ is more stringent than the topological connectedness of its support, and may depend on the choice of the exponent p, see [12] for more details.

Theorem 4.6. *Assume that μ is p-connected. Then the functional F^{hom} can be represented in the integral form*

$$F^{\mathrm{hom}}(u) = \int_\Omega f^{\mathrm{hom}}(\nabla u(x))\, dx\,, \qquad u \in H_0^{1,p}(\Omega)\,,$$

where the effective integrand f^{hom} can be computed, for every $z \in \mathbb{R}^n$, as the solution of the unit-cell problem

$$f^{\mathrm{hom}}(z) = \inf \left\{ \int_Y f(z + \nabla u(y)) \, d\mu \; : \; u \in C_{\sharp}^{\infty}(Y) \right\}$$

$$= \inf \left\{ \int_Y f_\mu(y, z + \nabla_\mu u(y)) \, d\mu \; : \; u \in H_{\mu,\sharp}^{1,p} \right\}, \tag{23}$$

being f_μ defined by (17). Moreover, we have the equality

$$F^{\mathrm{hom}} = \Gamma\text{-}\lim_{\delta \to 0} \left(\Gamma\text{-}\lim_{\epsilon \to 0} F_\epsilon^\delta \right). \tag{24}$$

Theorem 4.6 has been proved in [12] by means of a two-scale technique with respect to measures. Let us add a few comments on the statement. The first part extends to a much more general framework, a result which was known for special choices of the initial measure μ: for instance, when $\mu \llcorner Y$ is the Lebesgue measure over the complement in Y of a ball, we recover the homogenization of perforated domains [4]; also, the case when μ is concentrated on a periodic structure of codimension 1 had been previously studied in [2]. Since the p-connectedness assumption on μ involves no regularity or dimension condition on its support, Theorem 4.6 can be applied to the homogenization of very general structures of arbitrary codimension. Note that the practical computation of the effective energy density through (23) can be obtained in a straightforward way in many examples (in particular when μ is one-dimensional). The second part of the statement is a commutativity result. Here the Γ-limits in ϵ and δ are considered with respect to the convergence (21) and (22) respectively. Therefore, the equality (24) tells that the procedure usually employed in the literature, namely letting go to zero ϵ and δ in the order, may be reversed, following the shortcut of homogenizing the sequence $\{F_\epsilon\}$ associated to the measure $\mu = \lim_{\delta \to 0} \mu^\delta$.

Vector case. In this case, by effective energy J^{hom} associated to the sequence $\{J_\epsilon\}$, we do not simply mean the Γ-limit of J_ϵ. The reason is that, in several situations, such Γ-limit is not meaningful, as it may degenerate to zero even for very simple thin structures. As shown in [13], this is due to the behaviour of orthogonal displacements to the support of μ. Therefore, we set

$$J^{\mathrm{hom}} := \Gamma\text{-}\lim_{\delta \to 0} \left(\Gamma\text{-}\lim_{\epsilon \to 0} J_\epsilon^\delta \right).$$

We say that μ is *p-fat* if the following Poincaré-Korn inequality on "integer dilatations" of Y holds:

$$\exists C > 0 : \int_{kY} |u|^p \, d\mu \le C \, k^p \! \int_{kY} |e_\mu u|^p \, d\mu \; \forall k \in \mathbb{N}, \; \forall u \in \mathcal{D}_\mu^{1,p}(kY) : \int_{kY} u \, d\mu = 0,$$

where $\mathcal{D}_\mu^{1,p}(kY) := \left\{ u \in (L_{\mu,loc}^p)^n \; : \; u \text{ is } kY\text{-periodic and } u\psi \in \mathcal{D}_\mu^{1,p} \; \forall \psi \in \mathcal{D} \right\}$. The reason of the terminology "p-fat" is that the above condition is *not* satisfied by measures μ associated to thin structures, for instance by the one-dimensional Hausdorff measure over a system of periodic fibers in $\mathbb{R}^3$ (see [13], Section 4.2).

Theorem 4.7. *Assume that, for every $\delta > 0$, the measure μ^δ is p-fat, and assume that the quadratic form*

$$q(z) = \inf \left\{ \int_Y j(z + eu(y)) \, d\mu \; : \; u \in C_\sharp^\infty(Y; \mathbb{R}^n) \right\}$$

$$= \inf \left\{ \int_Y j_\mu(y, z + e_\mu u(y)) \, d\mu \; : \; u \in \mathcal{D}_{\mu,\sharp}^{1,p} \right\}$$

is positive definite on $\mathbb{R}_{\text{sym}}^{n^2}$, where $j_\mu(y, z) := \inf \left\{ j(z + \xi) \; : \; \xi \in [M_\mu(y)]^\perp \right\}$. Then the functional J^{hom} can be represented in the integral form

$$J^{\text{hom}}(u) = \int_\Omega j^{\text{hom}}(eu(x)) \, dx , \qquad u \in H_0^{1,p}(\Omega; \mathbb{R}^n) ,$$

where the effective integrand j^{hom} is given, for every $z \in \mathbb{R}_{\text{sym}}^{n^2}$, by $j^{\text{hom}}(z) = q(z)$.

We stress that, in full contrast with the scalar case, the δ-fattening approach is necessary to study the homogenization of vector problems on low-dimensional structures. Indeed, for thin elastic structures such that the approximating measures μ^δ are p-fat, Theorem 4.7 provides the nontrivial effective energy J^{hom}, whereas, letting δ tend to zero *before* making the homogenization procedure, one may find a vanishing Γ-limit. A more detailed study of this phenomenon can be found in [13].

Optimal bounds. A relevant application of the above homogenization results is the characterization of all effective materials which can be obtained by homogenization of a given conductor, when it is placed on an arbitrary structure of prescribed dimension in $\mathbb{R}^n$, in a prescribed quantity per unit cell. To give a rigorous formulation, fix a positive integer parameter $k \leq n$, and a real number $m > 0$. If μ varies in the class $\mathcal{M}_{k,m}$ of positive 2-connected periodic measures with mass m per unit cell and $\dim T_\mu = k$, and the initial energy density is given by $f(z) = |z|^2$, the effective integrand f^{hom} given by Theorem 4.6 turns out to be a quadratic form, represented by a matrix A_μ^{hom} whose spectrum must satisfy suitable bounds. Viceversa, any symmetric matrix satisfying such bounds can be obtained by homogenization on the given class. More precisely, we have the following G-closure type result [9].

Theorem 4.8. *Let $\mu \in \mathcal{M}_{k,m}$, let $f(z) = |z|^2$, and let $f^{\text{hom}} = A_\mu^{\text{hom}} z \cdot z$ be the homogenized integrand given by Theorem 4.6. Then the eigenvalues λ_i of the matrix A_μ^{hom}, $i = 1, \ldots, n$, satisfy the inequalities*

$$0 \leq \lambda_i \leq m , \qquad \sum_i \lambda_i \leq km . \tag{25}$$

Conversely, for every symmetric real matrix M whose eigenvalues λ_i satisfy (25), there exist a measure $\mu \in \mathcal{M}_{k,m}$ such that $M = A_\mu^{\text{hom}}$.

A central problem in the classical homogenization theory is the characterization of all the effective materials which can be obtained by homogenization of two given conductors, when they are placed in arbitrary subsets of the unit cell having a prescribed volume proportion (see [26] and references therein). Theorem 4.8 solves the limit problem in which one of the conductors disappears, and the other one, represented by the measure μ, concentrates on a structure of prescribed low-dimension. The extension of this result to the vector case turns out to be quite delicate: keeping the constraints on the dimension and on the mass, some bounds for the homogenized elasticity tensor have been proved in [9].

4.3. Minimum and control problems

As a model for variational problems involving curvatures, let us consider the following minimum problem for the Willmore functional:

$$\min\left\{\int_S \left(1 + |h(S)|^2\right) d\mathcal{H}^k \ : \ S \in \mathcal{A}(k, \Gamma, \eta)\right\} , \tag{26}$$

where S varies in the class $\mathcal{A}(k, \Gamma, \eta)$ of smooth k-dimensional manifolds in $\mathbb{R}^n$, having as a boundary a prescribed $(k-1)$-manifold Γ, with assigned conormal field η. More in general, one can consider optimal control problems where the control variable is represented by a manifold S and the cost functional depends on the solution of a state equation on S. For instance, for S varying in the same class as above, let us consider the model problem

$$\min\left\{\int_S \left(1 + j(x, u_S)\right) d\mathcal{H}^k \ : \ S \in \mathcal{A}(k, \Gamma, \eta) \ , \ -\Delta_S u_S = g\right\} , \tag{27}$$

where $j = j(x, u)$ is a convex Carathéodory integrand, Δ_S is the Laplace-Beltrami operator on S, and g is a prescribed continuous source term defined on $\mathbb{R}^n$. We can take for instance $j(x, u) := |u - u_0(x)|^2$ where z is a given data (inverse problem) or $j(x, u) := g(x)u$ (then $\int j(x, u_S)$ represents the compliance of the system).

Variational problems such as (26) or (27) in general do not admit a solution, basically because minimizing sequences, seen as indicatrix functions of sets, do not converge to the indicatrix function of an admissible manifold. Moreover, while a wide literature is available about optimal control problems on open subsets of $\mathbb{R}^n$, no general technique is known when the control variable is represented by a manifold. Therefore, in order to obtain existence results, it is interesting to give a weak formulation through the use or measures. In case of problems (26) and (27), it reads respectively as

$$\min\left\{\int \left(1 + |h(\mu)|^2\right) d\mu \ : \ \mu \in \mathcal{M}(k, \Gamma, \eta)\right\} , \tag{28}$$

$$\min\left\{\int \left(1 + j(x, u_\mu)\right) d\mu \ : \ \mu \in \mathcal{M}(k, \Gamma, \eta) \ , \ -\operatorname{div}((\nabla_\mu u)\mu) = g\mu\right\} , \tag{29}$$

where $\mathcal{M}(k, \Gamma, \eta)$ is the class of measures $\mu \in \mathcal{M}_{BC}$ such that $\dim T_\mu = k$, and $\partial\mu = \eta\mathcal{H}^{k-1}\llcorner\Gamma$. Note that $\mathcal{M}(k, \Gamma, \eta)$ is larger than the initial class $\mathcal{A}(k, \Gamma, \eta)$, because it contains also measures with nonconstant density, or with integer multiplicity different than 1. In this wider framework, we would obtain existence results applying the direct method of the Calculus of Variations with respect to the weak* convergence of measures. To that aim, we observe first that the class $\mathcal{M}(k,\Gamma,\eta)$, is not weakly closed, which would prevent the existence of a minimizer even for coercive and lower semicontinuous costs. This is due to the constraint on the dimension. Consider for instance the case when $n = 1$, $k = 0$, Γ is empty and $\eta = 0$. Then the measures $\mu_n := (1/n)\sum_{i=1}^{n} \delta_{i/n}$ belong to $\mathcal{M}(k,\Gamma,\eta)$, whereas this is not the case for their weak limit, which is given by the Lebesgue measure $\mathcal{L}^1$ over the interval $(0,1)$. Note that the densities $(1/n)$ of μ_n tend to zero as $n \to +\infty$. Actually, the class $\mathcal{M}(k,\Gamma,\eta)$ becomes weakly closed under the additional constraint of an uniform lower bound on the k-dimensional density. More precisely, consider the class $\mathcal{M}(k,\Gamma,\eta,\alpha)$ of measures $\mu \in \mathcal{M}(k,\Gamma,\eta)$ which satisfy for a fixed constant α the inequality

$$\theta_k(\mu,x) := \lim_{\rho \to 0} \frac{\mu(B_\rho(x))}{\rho^k} \geq \alpha > 0 \quad \text{for } \mu\text{-a.e. } x \, . \tag{30}$$

We stress that, for every $\mu \in \mathcal{M}_{BC}$ with $\dim T_\mu = k$, the limit defining $\theta_k(\mu,x)$ exists finite for μ-a.e. x by a monotonicity argument (see [7], Lemma 2.9).

Lemma 4.9. *The class $\mathcal{M}(k,\Gamma,\eta,\alpha)$ is weakly closed. Moreover, if a sequence $\{\mu_h\}$ in $\mathcal{M}(k,\Gamma,\eta,\alpha)$ converges weakly to μ and $\mathrm{spt}(\mu_h)$ is connected for every h, then the support of μ is also connected.*

In view of the above Lemma, the existence of a solution to problems (28)–(29) in the class $\mathcal{M}(k,\Gamma,\eta,\alpha)$ is gained (possibly adding a connectedness constraint on the support of admissible measures) if we can prove that the cost functionals are both coercive and lower semicontinuous with respect to the weak convergence of measures. Since the costs under consideration are both larger than the mass of μ, the coercivity is immediately satisfied, and we can focus attention on lower semicontinuity. It is well known that, since their integrands are convex functions of $h(\mu)$ and u_μ, the integral functionals in (28)–(29) are lower semicontinuous for the convergence of measures $h(\mu_n)\mu_n \rightharpoonup h(\mu)\mu$ and $u_{\mu_n}\mu_n \rightharpoonup u_\mu\mu$ respectively. We are thus reduced to ask whether, for a sequence $\{\mu_n\} \in \mathcal{M}(k,\Gamma,\eta,\alpha)$, we have the implications

$$\mu_n \rightharpoonup \mu \quad \Rightarrow \quad h(\mu_n)\mu_n \rightharpoonup h(\mu)\mu \text{ and } u_{\mu_n}\mu_n \rightharpoonup u_\mu\mu \, . \tag{31}$$

In the former case, we can give a positive answer. In fact, we have the following more general result [7], which relates the convergence of the curvatures to the limiting behaviour of the dimension of tangent spaces. For $\mu \in \mathcal{M}_{BC}$, we denote by $|H(\mu)|$ the total variation of the vector measure $H(\mu)$.

Theorem 4.10. *Let $\{\mu_n\} \subset \mathcal{M}_{BC}$, with $\mu_n \rightharpoonup \mu$ and $\sup_n |H(\mu_n)| < +\infty$; up to subsequences, denote by $f\mu$ the weak limit of $(\dim T_{\mu_n})\mu_n$. Then*

(i) *the dimension cannot decrease under passage to the limit (i.e. the inequality $\dim(T_\mu) \geq f$ holds μ-a.e.);*

(ii) *if the dimension does not increase (i.e. $\dim(T_\mu) = f$ μ-a.e.), then $\mu \in \mathcal{M}_{BC}$ and we have the weak convergence of the curvatures $H(\mu_n) \rightharpoonup H(\mu)$.*

As a corollary of Lemma 4.9 and Theorem 4.10, we obtain the existence of a solution for minimum problems involving curvatures such as problem (28) on the class $\mathcal{M}(k, \Gamma, \eta, \alpha)$.

The second convergence in (31) is much more delicate. Since the state equation we are considering is variational, namely it is the Euler-Lagrange equation of an integral functional, the required implication $\mu_n \rightharpoonup \mu \Rightarrow u_{\mu_n}\mu_n \rightharpoonup u_\mu\mu$ will be satisfied provided the energies $F_n(u) := \int |\nabla_{\mu_n} u|^p \, d\mu_n$ on the spaces $H^{1,p}_{\mu_n}$ Γ-converge to the energy $F(u) := \int |\nabla_\mu u|^p \, d\mu$ on $H^{1,p}_\mu$. The convergence adopted for functions u is the same as in Subsection 4.2, namely $u_n\mu_n \rightharpoonup u\mu$. The study of this problem, made in [8], has revealed more interesting and hard than it was expected. Indeed, we have seen that on the class $\mathcal{M}(k, \Gamma, \eta, \alpha)$ we have the convergence of the mean curvatures. This happens because the projecting operators converge as measures, namely $P_{\mu_n}\mu_n \rightharpoonup P_\mu\mu$ (whence, passing to the distributional divergence, $H(\mu_n) \rightharpoonup H(\mu)$). Actually a Convex Analysis argument allows to prove that the weak convergence for the projectors implies also a stronger convergence, which generalizes the one usually adopted for Young measures according to lemma below.

Lemma 4.11. *Let $\{\mu_n\}$ be any sequence of measures with $\mu_h \rightharpoonup \mu$. If $P_{\mu_n}\mu_n \rightharpoonup P_\mu\mu$, then we also have*

$$\lim_{n \to +\infty} \int \varphi(x, P_{\mu_n}(x)) \, d\mu_n = \int \varphi(x, P_\mu(x)) \, d\mu(x) , \qquad \forall \varphi \in C_0(\mathbb{R}^n \times \mathbb{R}^{n^2}) .$$

This result means that we have no oscillations for the sequence $\{P_{\mu_n}\mu_n\}$, and hence for the tangent spaces associated to μ_n: therefore, it is natural to expect a good asymptotic behaviour from energies defined on varying Sobolev spaces. In spite, a gap phenomenon may occur between the Γ-lim inf and the Γ-lim sup of the integrals F_n, due to the possible difference between the spaces $H^{1,p}_\mu$ and $W^{1,p}_\mu$ associated with the limit measure μ. The following result has been proved in [8] for a sequence $\{\mu_n\} \subset \mathcal{M}(k, \Gamma, \eta, \alpha)$, under the additional technical conditions $h(\mu_n) \in (L^\infty_{\mu_n})^n$ and $h(\mu_n) \perp T_{\mu_n}$ on the mean curvatures of μ_n. The Γ-convergence for the energies F_n on $H^{1,p}_{\mu_n}$ is intended in the sense specified above.

Theorem 4.12. *For $\{\mu_n\}$ and $\{F_n\}$ as above, we have*

(i) $\Gamma\text{-}\lim \sup(F_n) \leq F(u) := \int |\nabla_\mu u|^p \, d\mu , \quad u \in H^{1,p}_\mu;$

(ii) $\Gamma\text{-}\lim \inf(G_n) \geq G(u) := \int |D_\mu u|^p \, d\mu , \quad u \in W^{1,p}_\mu.$

Whenever $H^{1,p}_\mu \neq W^{1,p}_\mu$, we have $F \neq G$, and the Γ-limit of F_n may be either F or G according to the choice of the approximating sequence $\{\mu_n\}$. For instance, consider the example made in Subsection 3.2, when μ is given by the Lebesgue measure $\mathcal{L}^2$ on $\Omega = (0,1)^2 \cup (-1,0)^2$, and we have $F \neq G$ for all $p > 2$. If $\{\mu_n\}$ is an approximating sequence of the kind $\mu_n = \mathcal{L}^2 \llcorner \Omega_n$, the associated sequence $\{F_n\}$ Γ-converges to F if $\Omega_n \cap B_\rho(0,0)$ is connected for ρ sufficiently small, and to G otherwise. Similar examples may be produced even for measures with fixed boundary.

As a consequence of Theorem 4.12, we can prove by now existence results only for very special kinds of optimal control problems on truss-like structures, when the optimal measure μ must satisfy *a priori* the equality "$H = W$". An alternative approach could be adopted formulating the state equation on the weak Sobolev space: for measures with fixed boundary, we conjectured in [8] that the energies $G_n(u) := \int |D_{\mu_n} u|^p \, d\mu_n$ on $W_{\mu_n}^{1,p}$ Γ-converge to the functional G on the weak space $W_\mu^{1,p}$. The proof of this conjecture, which is concerned with a delicate approximation problem, is still open and would provide a powerful method for dealing with manifolds as controls.

References

[1] W. K. Allard, *First variation of a varifold*, Annals of Math., **95** (1972), 417–491.

[2] H. Attouch and G. Buttazzo, *Homogenization of reinforced periodic one-codimensional structures*, Ann. Scuola Norm. Sup. Pisa Cl. Sci., **14 (4)** (1987), 465–484.

[3] L. Ambrosio, N. Fusco and D. Pallara, Functions of Bounded Variation and Free Discontinuity Problems, Oxford University Press (2000).

[4] A. Bensoussan, J. L. Lions and G. Papanicolaou, Asymptotic Analysis for Periodic Structures, North-Holland, Amsterdam (1978).

[5] G. Bouchitté, G. Bellettini and I. Fragalà, *BV functions with respect to a measure and relaxation of metric integral functionals*, J. Convex Anal., **6 (2)** (1999), 349–366.

[6] G. Bouchitté and G. Buttazzo, *Characterization of optimal shapes and masses through Monge-Kantorovich equation*, J. Eur. Math. Soc. **3** (2001), 139–168.

[7] G. Bouchitté, G. Buttazzo and I. Fragalà, *Mean curvature of a measure and related variational problems*, Ann. Scuola Norm. Sup. Pisa Cl. Sci., **XXV (4)** (1997), 179–196.

[8] G. Bouchitté, G. Buttazzo and I. Fragalà, *Convergence of Sobolev spaces on varying manifolds*, to appear on J. Geom. Anal.

[9] G. Bouchitté, G. Buttazzo and I. Fragalà, *Bounds on the effective coefficients of homogenized low dimensional structures*, Preprint, (2001).

[10] G. Bouchitté, G. Buttazzo and P. Seppecher, *Energies with respect to a measure and applications to low dimensional structures*, Calc. Var. Partial Differential Equations **5** (1997), 37–54.

[11] G. Bouchitté, G. Buttazzo and P. Seppecher, *Optimization solutions via Monge-Kantorovich equation*, C. R. Acad. Sci. Paris, Série I **324** (1997), 1185–1191.

[12] G. Bouchitté and I. Fragalà, *Homogenization of thin structures by two-scale method with respect to measures*, SIAM J. Math. Anal., **32 (6)** (2001), 1198–1226.

[13] G. Bouchitté and I. Fragalà, *Homogenization of elastic thin structures: a measure-fattening approach*, Preprint, (2000).

[14] G. Bouchitté and I. Fragalà, *Second order energies on thin structures: variational theory and non-local effects*, Paper in preparation.

[15] G. Bouchitté and M. Valadier, *Integral representation of convex functionals on a space of measures*, J. Funct. Anal. **80** (1988), 398–420.

[16] G. Buttazzo, Semicontinuity, Relaxation, and Integral Representation in the Calculus of Variations, Pitman Research Notes in Mathematics Series, Longman Scientific & Technical, Harlow (1989).

[17] P. G. Ciarlet and H. Le Dret, *Justification de la condition aux limites d'encastrement d'une plaque par une méthode asymptotique*, C. R. Acad. Sci. Paris. Sér. I Math., **307 (20)** (1988), 1015–1018.

[18] D. Cioranescu and J. Saint Jean Paulin, Homogenization of Reticulated Structures. Applied Mathematical Sciences **136**, Springer Verlag, Berlin-Heidelberg-New York (1999).

[19] B. Dacorogna, Direct Methods in the Calculus of Variations, Applied Mathematical Sciences **78**, Springer-Verlag, Berlin-Heidelberg-New York (1988).

[20] G. Dal Maso, An Introduction to Γ-convergence, Birkhäuser, Boston (1993).

[21] H. Federer, Geometric Measure Theory, Springer-Verlag, Berlin-Heidelberg-New York (1969).

[22] H. Federer and W. Fleming, *Normal and integral currents*, Annals of Math., **72** (1960), 458–520.

[23] I. Fragalà, *Lower semicontinuity of μ-quasiconvex integrals*, Preprint, (2000).

[24] I. Fragalà and C. Mantegazza, *On some notions of tangent space to a measure*, Proc. Roy. Soc. Edinburgh, **129A** (1999), 331–342.

[25] M. Giaquinta, G. Modica and J. Soucek, Cartesian Currents in the Calculus of Variations, Springer-Verlag, Berlin-Heidelberg-New-York (1998).

[26] V. V. Jikov, S. M. Kozlov and O. A. Oleinik, Homogenization of Differential Operators and Integral Functionals, Springer-Verlag, Berlin-Heidelberg-New York (1994).

[27] H. Le Dret and A. Raoult, *The nonlinear membrane model as variational limit of nonlinear three dimensional elasticity*, J. Math. Pures Appl., **74** (1995), 549–578 .

[28] D. Percivale, *The variational methods for tensile structures,* Preprint, (1991).

[29] D. Preiss, *Geometry of measures on $\mathbb{R}^n$: distribution, rectifiability and densities*, Annals of Math., **125** (1987), 573–643.

[30] L. Simon, Lectures on Geometric Measure Theory, Proc. Centre for Math. Anal., Australian Nat. Univ., **3** (1983).

[31] M. Valadier, *Multiapplications mesurables à valeurs convexes compactes,* J. Math. Pures Appl., **50** (1971), 265–297.

Guy Bouchitté
Département de Mathématiques,
Université de Toulon et du Var,
F-83957 La Garde, Cedex
France
E-mail address: bouchitte@univ-tln.fr

Ilaria Fragalà
Dipartimento di Matematica – Politecnico,
Piazza Leonardo Da Vinci, 32
I-20133 Milano
Italy
E-mail address: fragala@mate.polimi.it

Progress in Nonlinear Differential Equations
and Their Applications, Vol. 51, 41–65
© 2002 Birkhäuser Verlag Basel/Switzerland

Optimal Transportation Problems with Free Dirichlet Regions

Giuseppe Buttazzo, Edouard Oudet and Eugene Stepanov

Abstract. A Dirichlet region for an optimal mass transportation problem is, roughly speaking, a zone in which the transportation cost is vanishing. We study the optimal transportation problem with an unknown Dirichlet region Σ which varies in the class of closed connected subsets having prescribed 1-dimensional Hausdorff measure. We show the existence of an optimal Σ_{opt} and study some of its geometrical properties. We also present numerical computations which show the shape of Σ_{opt} in some model examples.

1. Introduction

Optimal mass transportation problems received a lot of attention in the last years, among all, also for extensive connections with other fields such as shape optimization, fluid mechanics, partial differential equations, geometric measure theory (see [5, 4, 12, 13, 14]). Given two nonnegative measures f^+ and f^- over $\mathbf{R}^N$ the problem consists in the optimization of the cost of transporting f^+ into f^- by means of a *transport map* $T : \mathbf{R}^N \to \mathbf{R}^N$. More precisely, we say that T transports f^+ into f^- and call T a *transport map*, if $T_\# f^+ = f^-$ where $T_\#$ is the push-forward operator, that is

$$f^+\left(T^{-1}(B)\right) = f^-(B) \qquad \text{for every Borel set } B \subset \mathbf{R}^N. \tag{1}$$

Clearly, in this way, in order to have a nonempty class of admissible transport maps, we have to require that the two measures f^+ and f^- have the same mass.

For every transport map T the cost of transporting f^+ to f^- is defined by

$$J(T) = \int c\left(x, T(x)\right) df^+(x), \tag{2}$$

where $c(x, y)$ is a given continuous nonnegative function. The problem of optimal mass transportation is then

$$\min\left\{ J(T) \ : \ T \text{ transports } f^+ \text{ into } f^- \right\}. \tag{3}$$

Usually the cost density $c(x, y)$ is taken as a function of the Euclidean distance. In particular, one often chooses $c(x, y) = |x - y|^p$. The case $p = 1$ is then the classical Monge transportation problem and is related to several problems in

Received by the editors November 12, 2001.

shape optimization theory (see [4]). The case $p = 2$ is also widely studied for its applications in fluid mechanics, while the case $p < 1$, or more generally when c is a concave function, seems to be the most realistic for several applications, and has been studied in [7]. We quote as general surveys on mass transportation problems the book [12] as well as the monographs [6, 5, 1, 13, 14], where the reader may find all the details that here, for the sake of brevity, will be omitted.

When the measures f^+ and f^- may concentrate on lower dimensional sets, it may happen that no admissible transport maps exist. This is for instance the case, even if $N = 1$, when $f^+ = 2\delta_0$ and $f^- = \delta_1 + \delta_{-1}$. For this reason it is convenient to consider, the relaxed formulation of the problem due to Kantorovich, which uses instead of transport maps T, the so-called *transport plans*, which are nonnegative Borel measures γ on the product space $\mathbf{R}^N \times \mathbf{R}^N$ such that

$$\pi^+_\# \gamma = f^+, \qquad \pi^-_\# \gamma = f^-,$$

where π^+ and π^- are the projections of $\mathbf{R}^N \times \mathbf{R}^N$ on the first and second factors respectively. It is easy to see (cf. [1]) that a transport map T always induces a transport plan γ given by $\gamma = (\mathrm{Id} \times T)_\# f^+$. Conversely, every transport plan γ which is concentrated on a γ-measurable graph Γ is induced by a suitable transport map T. The cost of a transport plan γ is simply given by

$$J(\gamma) = \int c(x,y) \, d\gamma(x,y), \tag{4}$$

so that the optimal mass transportation problem becomes

$$\min \{ J(\gamma) \ : \ \gamma \text{ transport plan of } f^+ \text{ into } f^- \}. \tag{5}$$

Again we notice for the class of admissible transport plans to be nonempty, the measures f^+ and f^- must have the same mass.

Most often one considers the optimal mass transportation problem (5) constrained to a set $K \subset \mathbf{R}^N$. The latter represents a region to which the transportation process is confined. This simply means that all the geodesic paths along which the mass is carried have to remain into K. In what follows we assume that $K = \bar{\Omega}$ is the closure of a smooth connected bounded open set $\Omega \subset \mathbf{R}^N$. The function $c(x,y)$, which measures the cost to carry the mass from the point x to the point y, has then to take into account that, unless Ω is convex, the shape of Ω modifies the length of geodesic paths. In particular, the cost function of the form $c(|x - y|)$ has to be replaced by $c(d_\Omega(x,y))$ where d_Ω is the *geodesic distance* on Ω given by the formula

$$d_\Omega(x,y) := \inf \left\{ \int_0^1 |\alpha'(t)| \, dt \ : \ \alpha \in \mathrm{Lip}([0,1];\bar{\Omega}), \ \alpha(0) = x, \ \alpha(1) = y \right\}.$$

Furthermore, in applications (see for instance [4] for the relations with shape optimization problems) one often considers the presence of the so-called Dirichlet region $\Sigma \subset \bar{\Omega}$ (in the sequel assumed to be a closed set), which represents the zone where the cost of transportation vanishes. Heuristically it means that you are allowed to transport mass free of charge "along Σ". More formally, it means that the presence of Σ modifies the distance which governs the optimal mass transportation problem. In fact, setting

$$d_{\Omega,\Sigma}(x,y) \quad := \quad \inf\left\{d_\Omega(x,y) \wedge (d_\Omega(x,\xi_1) + d_\Omega(y,\xi_2)) \ : \ \xi_1,\xi_2 \in \Sigma\right\}$$

we obtain a semi-distance on $\bar\Omega$ which does not count the paths that both start and end in Σ. We also generalize the notion of a transport plan for the case of the presence of a nonempty Dirichlet region $\Sigma \subset \bar\Omega$, saying that a Borel measure γ over $\bar\Omega \times \bar\Omega$ is a transport plan of f^+ into f^-, if

$$\pi_\#^+ \gamma - \pi_\#^- \gamma = f^+ - f^- \text{ on } \bar\Omega \setminus \Sigma.$$

Plugging now the above semi-distance instead of d_Ω into the problem (5), we obtain the new optimal mass transportation problem [4]

$$\min\left\{ \int \phi\big(d_{\Omega,\Sigma}(x,y)\big)\, d\gamma(x,y) \ : \ \gamma \text{ transport plan of } f^+ \text{ into } f^-\right\}. \tag{6}$$

Note that for the latter problem, in view of the generalized definition of a transport plan, it is not necessary to require that f^+ and f^- have the same mass.

In this paper we are studying the optimization problem of finding the "best possible" Dirichlet region $\Sigma \subset \bar\Omega$ subject to certain constraints. Namely, we will call $MK(\Sigma)$ the minimum of the problem (6) and we will study the minimization of MK with respect to Σ. In the case $f^+ := \mathcal{L}^N \llcorner \Omega$ and $f^- = 0$ the functional MK reduces to the average distance functional

$$MK(\Sigma) = \int_\Omega \text{dist}_\Omega(x,\Sigma)\, dx,$$

where $\text{dist}_\Omega(x,\Sigma) := \inf_{y \in \Sigma} d_\Omega(x,y)$.

The natural constraints for Σ are as follows: Σ can vary in the class of closed subsets of $\bar\Omega$ with prescribed length (i.e. Hausdorff $\mathcal{H}^1$ measure) and with prescribed finite number of connected components. In fact, it is clear that if either of the constraints is dropped, then the infimum of this optimization problem in Σ is trivially zero. When $f^+ := \mathcal{L}^N \llcorner \Omega$, $f^- = 0$ (i.e. MK is just the average distance functional) and the length constraint on Σ is zero, the above problem turns out to be the problem of optimal location of a finite number of points in a set $\bar\Omega$. The latter has a lot of applications in economics and urban planning, but despite being extensively studied recently, still lacks a complete understanding of the qualitative properties of solutions. For a recent survey on this problem we refer the reader to [9]. In this paper we focus our attention on the case of nonzero length constraint, and, just for the sake of simplicity, the set Σ is required to be connected (it will be clear from our results that allowing a finite number of connected components will not change the qualitative properties of each of the components). We will show that an optimal Σ_{opt} exists and we study some geometrical properties of Σ_{opt} in the simplest situation when MK is reduced to the average distance functional. We also present some numerical computations which show the shape of Σ_{opt} in some model examples in order to justify some of our conjectures which still lack a rigorous proof.

2. Existence of optimal sets

Let $l \geq 0$ be fixed and let Ω be a bounded connected subset of $\mathbf{R}^N$ with a Lipschitz boundary. We also fix two non-negative measures f^+ and f^- on $\bar{\Omega}$ and consider the optimization problem

$$\min \; \left\{ MK(\Sigma) : \Sigma \subset \bar{\Omega} \text{ closed, connected, } \mathcal{H}^1(\Sigma) \leq l \right\} \qquad (7)$$

where the functional MK is defined in the introduction as the minimum value of problem (6). We have the following existence result.

Theorem 2.1. *Let the function ϕ appearing in (6) be continuous. Then the problem (7) admits a solution.*

Proof. Let a sequence $\{\Sigma_\nu\}_{\nu=1}^\infty$ of closed connected subsets of $\bar{\Omega}$ be a minimizing sequence for the functional MK, satisfying $\mathcal{H}^1(\Sigma_\nu) \leq l$ for all $\nu \in \mathbf{N}$. According to the Blaschke theorem (Theorem 4.4.6 of [3]) one has $\Sigma_\nu \to \Sigma$ in the sense of Hausdorff convergence up to a subsequence (not relabeled), while $\Sigma \subset \bar{\Omega}$ is still closed and connected. Moreover, in view of the Golab theorem (Theorem 4.4.7 of [3]) one also has $\mathcal{H}^1(\Sigma) \leq l$. Observing now that the Hausdorff convergence implies $d_\Omega(x, \Sigma_\nu) \to d_\Omega(x, \Sigma)$ for all $x \in \bar{\Omega}$, we obtain

$$d_{\Omega, \Sigma_\nu}(x, y) \to d_{\Omega, \Sigma}(x, y)$$

for all $(x, y) \in \bar{\Omega} \times \bar{\Omega}$. Moreover, since all d_{Ω, Σ_ν} are Lipschitz-continuous for the Euclidean distance with the same Lipschitz constant, then the convergence is actually uniform.

Let now γ_ν be the respective optimal transport plans, i.e.

$$MK(\Sigma_\nu) = \int_{\bar{\Omega} \times \bar{\Omega}} \phi(d_{\Omega, \Sigma_\nu}(x, y)) \, d\gamma_\nu(x, y),$$

while

$$\pi_\#^+ \gamma_\nu - \pi_\#^- \gamma_\nu = f^+ - f^- \text{ on } \bar{\Omega} \setminus \Sigma_\nu.$$

The sequence $\{\gamma_\nu(\bar{\Omega})\}_{\nu=1}^\infty$ can be assumed bounded, and hence up to a subsequence (again not relabeled) $\gamma_\nu \rightharpoonup \gamma$ $*$-weakly in the sense of measures, where γ is some positive Borel measure over $\bar{\Omega}$. Clearly then

$$\pi_\#^+ \gamma - \pi_\#^- \gamma = f^+ - f^- \text{ on } \bar{\Omega} \setminus \Sigma.$$

In fact, for every $\psi \in C_0(\bar{\Omega} \setminus \Sigma)$ one has

$$\int_{\bar{\Omega}} \psi \, d(\pi_\#^+ \gamma - \pi_\#^- \gamma) = \lim_\nu \int_{\bar{\Omega}} \psi \, d(\pi_\#^+ \gamma_\nu - \pi_\#^- \gamma_\nu) = \int_{\bar{\Omega}} \psi \, d(f^+ - f^-),$$

since every function with compact support in $\bar{\Omega} \setminus \Sigma$ has also compact support in $\bar{\Omega} \setminus \Sigma_\nu$ for sufficiently large $\nu \in \mathbf{N}$ (this follows from the convergence $\Sigma_\nu \to \Sigma$ in the sense of Hausdorff). At last it remains to observe that

$$MK(\Sigma) \leq \int_{\bar{\Omega} \times \bar{\Omega}} \phi(d_{\Omega, \Sigma}(x, y)) \, d\gamma(x, y) = \lim_\nu \int_{\bar{\Omega} \times \bar{\Omega}} \phi(d_{\Omega, \Sigma_\nu}(x, y)) \, d\gamma_\nu(x, y),$$

which shows that Σ is a minimizer of the problem. $\qquad \square$

Remark 2.2. A word-to-word restating of this proof shows even a formally slightly more general result, namely, that the functional MK attains a minimum even over a class of closed subsets $\Sigma \subset \bar{\Omega}$ with $\mathcal{H}^1(\Sigma) \leq l$ and having a fixed prescribed number of connected components. The latter case includes also the situation $l = 0$ but the number of connected components is greater than one. If $f^+ := \mathcal{L}^N \llcorner \Omega$ and $f^- = 0$, the functional MK reduces to the average distance functional, and (7) is just the problem of optimal location of a finite number of points in $\bar{\Omega}$ (see [9]).

3. Qualitative properties of optimal sets

In this section we consider some qualitative properties every optimal solution Σ_{opt} to the problem (7) has to fulfill. We present here some problems together with some conjectures which we believe are true. We also present some numerical approximations of Σ_{opt} in different particular situations as well as proofs of some results which, though sometimes weaker than our expectations, still induce to think that the conjectures we formulate most probably hold true.

When formulating the problems below we assume that $\Omega \subset \mathbf{R}^N$ is a bounded connected open set with Lipschitz boundary and that the measures f^+ and f^- are absolutely continuous with respect to the Lebesgue measure and different from each other. For simplicity we consider only the case of the cost function $\phi(t) = t$, though most of the questions below could also be raised for more general cost functionals.

Problem 3.1 (Regularity). *Study the regularity properties of the solutions Σ_{opt}. We actually expect that Σ_{opt} are piecewise smooth, i.e. made of a finite number of smooth curves connected through a finite number of singular points.*

Problem 3.2 (Absence of loops). *Study the topological properties of the solutions Σ_{opt}. We expect that Σ_{opt} do not form closed loops in Ω. When the dimension N is equal to 2 this can be expressed by saying that $\mathbf{R}^2 \setminus \Sigma_{opt}$ is connected.*

Problem 3.3 (Triple points). *Study the nature of the singular points mentioned in Problem 3.1. We expect that they can only be triple points, that is points where three curves meet, with angles of 120 degrees.*

Problem 3.4 (Distance from the boundary). *Study the cases when the optimal solutions Σ_{opt} do not touch the boundary $\partial\Omega$. We expect that this occurs at least when Ω is convex.*

Problem 3.5 (Behavior for small lengths). *Study the asymptotic behaviour of Σ_{opt} as $l \to 0$. In particular we expect that for l small enough Σ_{opt} is a smooth curve without singular points. More in general, it would be interesting to obtain an estimate of the number of singular points in terms of the length of Σ_{opt}.*

Problem 3.6 (Behavior for large lengths). *Study the asymptotic behaviour of Σ_{opt} as $l \to +\infty$. It is not difficult to see that the value V_l of the optimization problem (7) vanishes as $l \to +\infty$. It would be interesting to evaluate the order of the* *vanishing quantity V_l. Moreover, once we estimate that $V_l = O(l^{-\beta})$ for some $\beta > 0$ as $l \to +\infty$, it is interesting to study the Γ-limit as $l \to +\infty$ of the rescaled functionals*

$$G_l(\Sigma) = \frac{1}{l^{\beta}} MK(\Sigma)$$

with respect to the convergence

$$\Sigma_l \to \lambda \quad \Leftrightarrow \quad \frac{1}{l}\mathcal{H}^1 \llcorner \Sigma_l \to \lambda \quad weakly^* \text{ in the sense of measures.}$$

In particular, if Σ_l are optimal configurations of length l, it is interesting to study the asymptotic behaviour of Σ_l as well as their limit λ in the sense above.

We now start to develop the program introduced above by considering the simpler situation when Ω is convex, f^+ is the Lebesgue measure on Ω, $f^- = 0$. Then the functional $MK(\Sigma)$ reduces to the average distance functional and d_Ω to the Euclidean distance, so that problem (7) becomes

$$\min\left\{\int_\Omega \text{dist}\,(x, \Sigma)\,dx : \Sigma \subset \bar{\Omega} \text{ closed, connected, } \mathcal{H}^1(\Sigma) \leq l\right\}. \qquad (8)$$

3.1. The problem in a unit disk of $\mathbf{R}^2$

We start by considering the case when Ω is the unit disc in $\mathbf{R}^2$. The first guess for Σ_{opt} in the case l is sufficiently small could be a circumference centered at the origin, which is however ruled out by Theorem 3.10. A second guess for Σ_{opt}, always in the case of a sufficiently small l, would be a segment centered at the origin. This is again excluded by the proposition below, the proof of which can be found in Appendix A.

Proposition 3.7. *There exists $l_0 > 0$ such that for all $l \leq l_0$ the centered segment of length l is not optimal for problem (8).*

We will now discuss the numerical approximations of the optimal set in a unit disc of $\mathbf{R}^2$. We will limit ourselves to presenting the results and ideas of the methods rather than the technical aspects of our algorithms. The latter will be described in details in [10]. Two cases will be subject of our numerical study:

(i) Σ is a set consisting of finite prescribed number of points (location problem);

(ii) Σ is a compact connected set with prescribed length.

For each of the above constraints we use a different numerical approach.

3.1.1. OPTIMAL LOCATION OF A FINITE NUMBER OF POINTS.

Given $n \in \mathbf{N}$, we are looking for an optimal n-point set Σ_n, which minimizes the quantity (8). If $n = 1$, it is not difficult to prove that the only minimizer is the center of the disk. When $n \geq 2$ is not too large we could guess that the optimal set Σ_n is given by the vertices of a centered regular polygon.

To approximate numerically Σ_n for $n > 1$, we use the classical finite difference method. As underlined in [8], this is a reasonable way to solve design optimization problems with few parameters. We present in Figure 1 two pictures obtained by this process. The first one shows that for $n = 5$ the optimal set Σ_5 seems to be distributed on a regular centered polygon. The same situation occurs for $n = 2, 3, 4$. The second image represents the case of $n = 6$, and one can observe that the center of the disk is one of

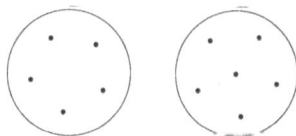

FIGURE 1. Optimal locations of 5 and 6 points in a disk.

3.1.2. OPTIMAL COMPACT CONNECTED SET WITH PRESCRIBED LENGTH. We are now looking for an optimal set Σ_{opt} among all compact connected sets of the disk with one-dimensional Hausdorff measure not exceeding l. This situation is definitely more complex from a numerical point of view than the previous one. The main problem is to build a process which is able to identify the topology of Σ_l. Moreover, it is intuitively clear that this problem has a great number of local minima. Last but not least, the length constraint is rather difficult to handle numerically. Considering these difficulties, it seems natural to use an Evolutionary Algorithm (EAs) with an adaptive penalty method. We will not present here the theory of EAs but we refer the interested reader to [11] for an introduction to Adaptive methods in EAs. Further numerical details like the representation of Σ_{opt}, the cost function, the adaptive penalty and different test cases will be described more accurately in [10].

In Figure 2 one can see the values $V(l)$ obtained by the chosen numerical method as a function of length constraint l. They are compared to numerical evaluations of (8) for some simple sets like a circumference, a regular cross (of two perpendicular intervals), a regular trisection (i.e. 3 equal intervals joined at one of

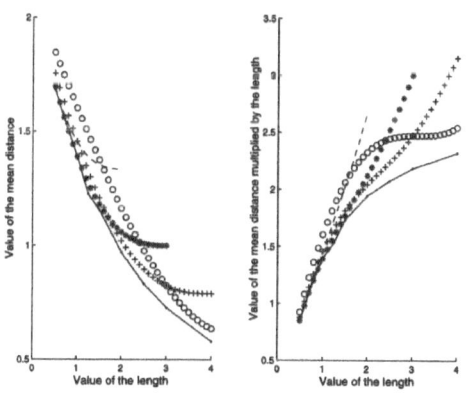

FIGURE 2

their endpoints at the angle of 120 degrees each) and a segment, all centered at the origin. In this figure

- o   stands for the cirumference centered at the origin,
- +   stands for the centered perpendicular cross,
- *   stands for the regular trisection,
- - -   stands for the segment,
- —   stands for the numerical approximations of optimal sets.

In the same figure we present a graph of $lV(l)$ as a function of l (in fact, in Theorem 3.16 we will show that the quantity $lV(l)$ for the optimal set is bounded as $l \to +\infty$).

In Figures 3–5 the numerical approximations of optimal sets in a disc for different lengths are shown. Perhaps somewhat unexpected is the fact that the number of singular points does not seem to be an increasing function of the prescribed length. More numerical computations are presented in Appendix B.

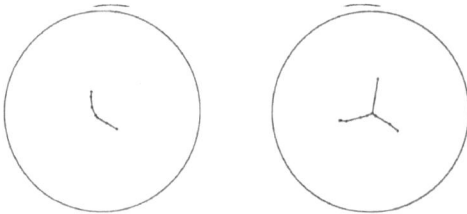

FIGURE 3. Optimal sets of length 0.5 and 1 in a unit disk

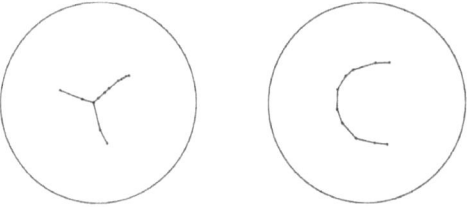

FIGURE 4. Optimal sets of length 1.25 and 1.5 in a unit disk

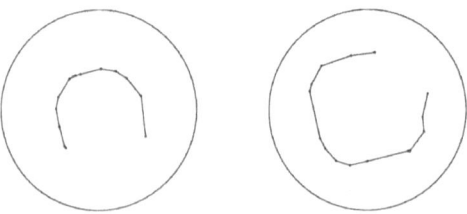

FIGURE 5. Optimal sets of length 2 and 3 in a unit disk

3.2. Singular points

From now on we present some simple results on the qualitiative properties of the optimal set Σ_{opt} in a generic closed convex set $\bar{\Omega}$. The results we present here do not completely answer the questions raised at the beginning of this section, but rather indicate what kind of strong results one can reasonably expect.

We start from a very simple proposition which gives a partial answer to Problem 3.3, restricting, roughly speaking, the possible singularities of the optimal set to only triple points, i.e. points where three curves meet with angles of 120 degrees. In order not to overburden the paper with technical details we just formulate the proposition in its simplest way, namely, we assert that a cross can never be an optimal set in $\bar{\Omega} \subset \mathbf{R}^2$. We call a cross the union of two mutually perpendicular closed intervals intersecting in a point which is internal for both.

Proposition 3.8. *Let $N = 2$. Then a cross is not an optimal set.*

Proof. Assume the contrary, i.e. that Σ_{opt} is a cross, and assume without loss of generality that its center (i.e. the intersection point of two intervals) is the origin of coordinate system.

STEP 1. For every sufficiently small $\varepsilon > 0$ the set $D_\varepsilon := \Sigma_{opt} \cap \partial B_\varepsilon(0)$ consists of exactly 4 points. Denote by $S_4(D_\varepsilon) \subset B_\varepsilon(0)$ a set of minimum length in the ball $B_\varepsilon(0)$ which connects the all the four points of D_ε as in Figure 6 (we will call it a *Steiner connection* of these points) as in Figure 6. Observe now that

$$\mathcal{H}^1(\Sigma_{opt} \cap B_\varepsilon(0)) - \mathcal{H}^1(S_4(D_\varepsilon)) \geq C\varepsilon$$

for some $C > 0$ (here and below the value of the constant C may vary from line to line). In fact, to show this estimate, it is enough to prove ist rescaled version

$$\mathcal{H}^1((1/\varepsilon)\Sigma_{opt} \cap B_1(0)) - \mathcal{H}^1(S_4(D_1)) \geq C,$$

which follows from the direct computation

$$\mathcal{H}^1((1/\varepsilon)\Sigma_{opt} \cap B_1(0)) = 4 \text{ and } \mathcal{H}^1(S_4(D_1)) = \sqrt{2}(\sqrt{3}+1) < 4.$$

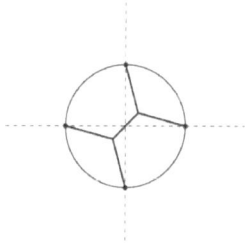

FIGURE 6. Steiner connection

STEP 2. Let now Σ_ε be Σ_{opt} outside of $B_\varepsilon(0)$ and $S_4(D_\varepsilon)$ inside $B_\varepsilon(0)$. Clearly then

$$\text{dist}\,(x, \Sigma_\varepsilon) \leq \text{dist}\,(x, \Sigma_{opt}) + 2\varepsilon$$

for all $x \in \bar{\Omega}$. Moreover, let Λ_ε be the set of points whose projection on Σ_{opt} is different from the projection on Σ_ε. It is easy to see that Λ_ε is contained inside a square centered at the origin of size ε. Therefore

$$\int_\Omega \text{dist}\,(x, \Sigma_\varepsilon)\, dx \leq \int_\Omega \text{dist}\,(x, \Sigma_{opt})\, dx + C\varepsilon^2 \qquad (9)$$

According to Step 1, we have however at least the additional length $C\varepsilon$ to use in order to decrease the functional. This is achieved by using Lemma 3.12 below. Namely, according to this lemma we can attach a segment of this length to Σ_ε obtaining a set Σ'_ε so that

$$\int_\Omega \text{dist}\,(x, \Sigma'_\varepsilon)\, dx \leq \int_\Omega \text{dist}\,(x, \Sigma_\varepsilon)\, dx - C\varepsilon^{3/2}. \qquad (10)$$

Now, in view of (9) and (10), one gets for sufficiently small $\varepsilon > 0$ the estimate

$$\int_\Omega \text{dist}\,(x, \Sigma'_\varepsilon)\, dx \leq \int_\Omega \text{dist}\,(x, \Sigma_{opt})\, dx - C\varepsilon^{3/2},$$

for some $C > 0$, which contradicts the optimality of Σ_{opt}. □

Remark 3.9. Arguments similar to the ones used in the above proof allow to show that our conjecture stated in Problem 3.3 is true at least if Σ_{opt} is piecewise sufficiently smooth.

3.3. Absence of loops

We present here a result giving a partial answer to Problem 3.2.

Theorem 3.10. Let $N = 2$. Then for $\mathcal{H}^1$-a.e. $x \in \Sigma_{opt}$ and for all sufficiently small $\varepsilon > 0$ the set $\Sigma_{opt} \setminus B_\varepsilon(x)$ is disconnected.

Proof. Suppose the contrary, namely that the set

$$A := \{x \in \Sigma_{opt} : \exists \{\varepsilon_\nu\}_{\nu=1}^\infty, \varepsilon_\nu \searrow 0 \text{ such that } \Sigma_{opt} \setminus B_{\varepsilon_\nu}(x) \text{ is connected}\}$$

has positive length, i.e. $\mathcal{H}^1(A) > 0$. Further on we omit the reference to the index ν writing always ε instead of ε_ν. Let $\tilde{A} \subset A$ stand for the set of density points of Σ_{opt}, i.e. such that for every $x \in \tilde{A}$ one has

$$\lim_{\varepsilon \to 0^+} \frac{\mathcal{H}^1(\Sigma_{opt} \cap B_\varepsilon(x))}{2\varepsilon} = 1.$$

Since Σ_{opt} is $(\mathcal{H}^1, 1)$-rectifiable, then $\mathcal{H}^1(\tilde{A}) = \mathcal{H}^1(A)$ by Besicovitch-Marstrand-Mattila theorem (Theorem 2.63 from [2]).

For every $x \in \tilde{A}$ and $\varepsilon > 0$ let $T(x, \varepsilon)$ stand for the union of transport rays of the Monge-Kantorovich problem of transporting $\mathcal{L}^N \llcorner \Omega$ to its projection over Σ which end at $\Sigma_{opt} \cap B_\varepsilon(x)$. Set $\Sigma_\varepsilon(x) := \Sigma_{opt} \setminus B_\varepsilon(x)$. Clearly, according to our assumption, $\Sigma_\varepsilon(x)$ is still closed and connected, and, of course, satisfies the length constraint since $\mathcal{H}^1(\Sigma_\varepsilon(x)) \le \mathcal{H}^1(\Sigma_{opt})$. The following estimate is valid

$$\int_\Omega \text{dist}\,(z, \Sigma_\varepsilon(x))\,dz$$

$$= \int_{\Omega \setminus T(x,\varepsilon)} \text{dist}\,(z, \Sigma_\varepsilon(x))\,dz + \int_{T(x,\varepsilon)} \text{dist}\,(z, \Sigma_\varepsilon(x))\,dz$$

$$\le \int_{\Omega \setminus T(x,\varepsilon)} \text{dist}\,(z, \Sigma_{opt})\,dz + \int_{T(x,\varepsilon)} (\text{dist}\,(z, \Sigma_{opt}) + \varepsilon)\,dz$$

$$= \int_\Omega \text{dist}\,(z, \Sigma_{opt})\,dz + \varepsilon \mathcal{L}^N(T(x,\varepsilon)).$$

One also has

$$\mathcal{L}^N(T(x,\varepsilon)) = \psi(B_\varepsilon(x)),$$

where ψ stands for the projection of $\mathcal{L}^N$ to Σ_{opt}. But

$$\limsup_{\varepsilon \to 0+} \frac{\psi(B_\varepsilon(x))}{\varepsilon} < +\infty$$

for $\mathcal{H}^1$-a.e. $x \in \tilde{A}$, which implies

$$\mathcal{L}^N(T(x,\varepsilon)) \le C\varepsilon$$

for some $C = C(x) > 0$ depending on x and for sufficiently small $\varepsilon > 0$. Summing up, for $\mathcal{H}^1$-a.e. $x \in \tilde{A}$ one has

$$\int_\Omega \text{dist}\,(z, \Sigma_\varepsilon(x))\,dz \le \int_\Omega \text{dist}\,(z, \Sigma_{opt})\,dz + C\varepsilon^2 \qquad (11)$$

once $\varepsilon > 0$ is sufficiently small.

Let now

$$l(\varepsilon) := \mathcal{H}^1(\Sigma_{opt}) - \mathcal{H}^1(\Sigma_\varepsilon(x)),$$

where $x \in \tilde{A}$ is such that (11) holds. Since we chose $x \in \tilde{A}$, one has $l(\varepsilon) = C\varepsilon + o(\varepsilon)$ for $\varepsilon \to 0$ for some $C = C(x) > 0$. Applying now Lemma 3.12 with $\Sigma_\varepsilon(x)$ instead of Σ and $l(\varepsilon)$ instead of ε to find a closed connected set $\Sigma' \subset \bar{\Omega}$ satisfying $\mathcal{H}^1(\Sigma') = \mathcal{H}^1(\Sigma) + l(\varepsilon)$ and

$$\int_\Omega \text{dist}\,(x, \Sigma')\,dx \le \int_\Omega \text{dist}\,(x, \Sigma_\varepsilon(x))\,dx - C\varepsilon^{3/2}$$

for some $C > 0$ as $\varepsilon \to 0^+$. Using (11), one arrives at the estimate

$$\int_\Omega \text{dist}\,(x, \Sigma')\,dx \le \int_\Omega \text{dist}\,(z, \Sigma_{opt})\,d(z) - C\varepsilon^{3/2}$$

for sufficiently small $\varepsilon > 0$, which gives a contradiction with the optimality of Σ_{opt}. $\square$

Remark 3.11. For a piecewise smooth Σ_{opt} the conclusion of Theorem 3.10 implies the absence of loops indicated in Problem 3.2. In fact in this case $\mathbf{R}^2 \setminus \Sigma_{opt}$ is connected.

Lemma 3.12. *Let $N = 2$ and $\Sigma \subset \bar{\Omega}$ be a compact connected set such that $\Sigma \cap \Omega \neq \emptyset$. Then there is a constant $C > 0$ such that for all $\varepsilon > 0$ there is a segment S_ε of length ε such that the following conditions hold:*

- *the set $\Sigma_\varepsilon := \Sigma \cup S_\varepsilon$ is connected;*
- *the inequality*

$$\int_\Omega \text{dist}\,(z, \Sigma_\varepsilon)\,dz \leq \int_\Omega \text{dist}\,(z, \Sigma)\,dz - C\varepsilon^{3/2}$$

is true for sufficiently small $\varepsilon > 0$.

Proof. Let $A \in \Sigma \cap \Omega$ and $O \in \Omega \setminus \Sigma$ such that A is a projection of O on Σ. Without loss of generality consider the coordinate axes to be positioned so that O is the origin of the coordinate system and the following conditions are satisfied:

- $B_1(0) \subset \Omega$;
- $\Sigma \cap B_1(0) = \emptyset$ where $B_1(0)$ stands for the unit open disc;
- $A := (0, -1) \in \Sigma \cap \bar{B}_1(0)$.

Let $A_\varepsilon := (0, -1 + \varepsilon)$, $S_\varepsilon := [A, A_\varepsilon]$ stand for the closed interval with endpoints A and A_ε and $\Sigma_\varepsilon := \Sigma \cup S_\varepsilon$. Consider the set

$$\Lambda_\varepsilon := \{z = (x, y) \in B_1(0) : y \leq 0,\ d(z, A_\varepsilon) - \text{dist}\,(z, \partial B_1(0)) \leq -\varepsilon/4\}.$$

If we are able to prove that $\mathcal{L}^N(\Lambda_\varepsilon) \geq C\varepsilon^{1/2}$ for some $C > 0$, then the proof will be finished. Indeed, for $z \in \Lambda_\varepsilon$ one has

$$\text{dist}\,(z, \Sigma_\varepsilon) \leq d(z, A_\varepsilon) \leq \text{dist}\,(z, \partial B_1(0)) - \varepsilon/4 \leq \text{dist}\,(z, \Sigma) - \varepsilon/4,$$

and hence

$$\int_\Omega \text{dist}\,(z, \Sigma_\varepsilon)\,dz - \int_\Omega \text{dist}\,(z, \Sigma)\,dz \leq -\varepsilon\mathcal{L}^N(\Lambda_\varepsilon)/2 \leq -C\varepsilon^{3/2}$$

It remains therefore to estimate $\mathcal{L}^N(\Lambda_\varepsilon)$ from below. Let $0 < k < 1/2$ and

$$\Pi_{k,\varepsilon} := \left\{z = (x, y) \in B_1(0) : |y + 1/2| \leq (1 - 4k^2)^{1/2}/2,\ |x| \leq k\varepsilon^{1/2}\right\}.$$

For every $(x, y) \in \Pi_{k,\varepsilon}$, we have

$$
\begin{aligned}
d(z, A_\varepsilon) - \text{dist}\,(z, \partial B_1(0)) &\leq \left(x^2 + (y + 1 - \varepsilon)^2\right)^{1/2} - 1 + \left(x^2 + y^2\right)^{1/2} \\
&\leq \left(k^2\varepsilon + (y + 1 - \varepsilon)^2\right)^{1/2} - 1 + \left(k^2\varepsilon + y^2\right)^{1/2} \\
&\leq -1 + |y| + k^2\varepsilon/2|y| + (y + 1) + \\
&\quad\ \varepsilon\left(k^2 - 2(y + 1)\right)/2(y + 1) + \alpha\varepsilon^2 \\
&= \varepsilon\left(k^2/(y + 1) - 2 + k^2/|y|\right)/2 + \alpha\varepsilon^2,
\end{aligned}
$$

where $\alpha = \alpha(k) > 0$ is some constant. It is easy to verify that since $z \in \Pi_{k,\varepsilon}$, then

$$k^2/(y + 1) - 2 + k^2/|y| \leq -1,$$

and hence

$$d(z, A_\varepsilon) - \text{dist}\,(z, \partial B_1(0)) \leq -\varepsilon/4$$

for sufficiently small $\varepsilon > 0$, which means $\Pi_{k,\varepsilon} \subset \Lambda_{\varepsilon}$ for such ε. Thus

$$\mathcal{L}^N(\Lambda_{\varepsilon}) \geq \mathcal{L}^N(\Pi_{k,\varepsilon}) \geq C\varepsilon^{1/2},$$

for sufficiently small $\varepsilon > 0$, which concludes the proof.   □

3.4. The optimal set and the boundary

We are able to give now a partial answer to Problem 3.4. It only says that the intersection of Σ_{opt} with $\partial\Omega$ cannot have a positive $\mathcal{H}^1$ measure. Moreover, it is only proven when $N = 2$ and for $\Omega \subset \mathbf{R}^2$ convex, with sufficiently regular boundary having everywhere positive curvature.

Theorem 3.13. *Let $\Omega \subset \mathbf{R}^2$ be a convex set with a C^2 boundary having everyhere positive curvature and Σ_{opt} be a solution to problem (8). Then $\mathcal{H}^1(\Sigma_{opt} \cap \partial\Omega) = 0$.*

Proof. Suppose the contrary, i.e. $\mathcal{H}^1(\Sigma_{opt} \cap \partial\Omega) =: \alpha > 0$. Consider then for every $\varepsilon > 0$ the set

$$\Omega_{\varepsilon} := \{x \in \Omega : \text{dist}\,(x, \partial\Omega) > \varepsilon\}$$

and let p_{ε} stand for the projection map on the closed convex set $\bar{\Omega}_{\varepsilon}$, namely,

$$p_{\varepsilon}(x) := x + (\varepsilon - d(x))^+ \nabla d(x),$$

where $d(x) := \text{dist}\,(x, \partial\Omega)$ and $(\cdot)^+$ stands for the positive part function. Let $\Sigma_{\varepsilon} := p_{\varepsilon}(\Sigma_{opt})$. Clearly, Σ_{ε} is still connected and compact.

STEP 1. Let us estimate from below

$$\mathcal{H}^1(\Sigma_{opt}) - \mathcal{H}^1(\Sigma_{\varepsilon}).$$

This difference is nonnegative since p_{ε} is Lipschitz continuous with constant one. To make a more precise estimate, we note that

$$\nabla p_{\varepsilon}(x) := Id + (\varepsilon - d(x))^+ \nabla^2 d(x) - \nabla d(x) \otimes \nabla d(x), \text{ for } 0 \leq d(x) < \varepsilon,$$

where Id stands for the identity matrix. In particular, for $x \in \partial\Omega$ one has

$$\nabla p_{\varepsilon}(x) := T + \varepsilon\nabla^2 d(x) \text{ where } T := Id - \nabla d(x) \otimes \nabla d(x).$$

A simple calculation shows that $(Tz, z) = |z|^2 - |z_{\nu}|^2 \leq |z|^2$, where z_{ν} stands for the projection of z on the direction of the normal $\nu := \nabla d(x)$ to the boundary $\partial\Omega$ at the point x. Since we assumed $\partial\Omega$ to be of class C^2, then $\nabla^2 d$ is continuous over $\partial\Omega$, and, moreover, since the curvature of $\partial\Omega$ is supposed to be strictly positive, then the matrix $-\nabla^2 d(x)$ is positive definite along the directions tangential to $\partial\Omega$. Namely, there is a $K > 0$ such that $-(\nabla^2 d(x)z_{\tau}, z_{\tau}) \geq K|z_{\tau}|^2$ for all $x \in \partial\Omega$, where z_{τ} stands for the projection of z on the tangent space to the boundary $\partial\Omega$ at the point x. Also, clearly $\nabla p_{\varepsilon}(x)z_{\nu} = 0$ for every $z \in \mathbf{R}^N$. Summing up, we have

$$|\nabla p_{\varepsilon}(x)z| = |\nabla p_{\varepsilon}(x)z_{\tau}| \leq (1 - K\varepsilon)|z_{\tau}| \leq (1 - K\varepsilon)|z|$$

for all $x \in \partial\Omega$.

Without loss of generality we may suppose that $\partial\Omega$ is parametrized by a curve and let $\gamma\colon [0, L] \to \partial\Omega$ be the arc-length parametrization of $\partial\Omega$. Then, setting $e := \Sigma_{opt} \cap \partial\Omega$, we get by the area formula

$$
\begin{aligned}
\mathcal{H}^1(p_\varepsilon(e)) &= \int_{\gamma^{-1}(e)} |\nabla p_\varepsilon(\gamma(t))\gamma'(t)|\, dt \\
&\leq (1 - K\varepsilon) \int_{\gamma^{-1}(e)} |\gamma'(t)|\, dt \\
&= (1 - K\varepsilon)\mathcal{H}^1(e).
\end{aligned}
$$

Hence we finally arrive at the estimate

$$
\mathcal{H}^1(\Sigma_{opt}) - \mathcal{H}^1(\Sigma_\varepsilon) \geq \mathcal{H}^1(e) - \mathcal{H}^1(p_\varepsilon(e)) \geq K\alpha\varepsilon.
$$

STEP 2. Clearly,

$$
\int_\Omega \operatorname{dist}(z, \Sigma_\varepsilon)\, dz \geq \int_\Omega \operatorname{dist}(z, \Sigma_{opt})\, dz.
$$

We now estimate more precisely the difference between the two integrals. For this purpose consider a point of $z \in \Omega$ for which

$$
\operatorname{dist}(z, \Sigma_\varepsilon) > \operatorname{dist}(z, \Sigma_{opt}). \tag{12}
$$

Denote by z_0 an arbitrary projection of z to Σ_{opt}. Then obviously $z_0 \in \bar\Omega \setminus \Omega_\varepsilon$. We claim first that $z \in \bar\Omega \setminus \Omega_\varepsilon$. In fact, suppose the contrary, and let $z_{0\varepsilon}$ stand for the projection fo z_0 to $\partial\Omega_\varepsilon$. Since Ω_ε is convex, then z belongs to a half-plane bounded by a tangent line to Ω_ε at $z_{0\varepsilon}$, and hence also to the half-plane bounded by a line passing through the center of the segment $[z_0, z_{0\varepsilon}]$ perpendicular to the latter (see Figure 7). This implies $d(z, :$ re

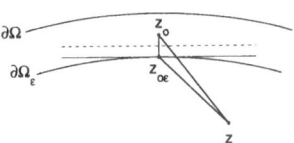

FIGURE 7

$$
\operatorname{dist}(z, \Sigma_\varepsilon) \leq d(z, z_{0\varepsilon}) < d(z, z_0) = \operatorname{dist}(z, \Sigma_{opt})
$$

which contradicts the assumption (12).

Let now z_ε stand for a projection of z on $\partial\Omega_\varepsilon$. One has then

$$
\operatorname{dist}(z, \Sigma_\varepsilon) - \operatorname{dist}(z, \Sigma_{opt}) = d(z, z_\varepsilon) - d(z, z_0) \leq d(z, z_\varepsilon) \leq \varepsilon.
$$

With the above estimate we get

$$
\begin{aligned}
\int_\Omega \operatorname{dist}(z, \Sigma_\varepsilon)\, dz - \int_\Omega \operatorname{dist}(z, \Sigma_{opt})\, dz &\leq \int_{\Omega\setminus\Omega_\varepsilon} (\operatorname{dist}(z, \Sigma_\varepsilon) - \operatorname{dist}(z, \Sigma_{opt}))\, dz \\
&\leq \varepsilon \mathcal{L}^2(\Omega\setminus\Omega_\varepsilon) \\
&= C\varepsilon^2 + o(\varepsilon^2)
\end{aligned} \tag{13}
$$

for some $C \geq 0$ as $\varepsilon \to 0^+$.

STEP 3. Consider $O \in \Omega_\varepsilon \backslash \Sigma_\varepsilon$ and its arbitrary projection O_ε to Σ_ε. We define

$$\Sigma_\varepsilon' := \Sigma_\varepsilon \cup S_\varepsilon$$

where S_ε is a segment of length $\mathcal{H}^1(\Sigma_{opt}) - \mathcal{H}^1(\Sigma_\varepsilon)$ starting at $p_\varepsilon(O)$ and pointing to O. In view of Lemma 3.12 one has

$$\int_\Omega \text{dist}(z, \Sigma_\varepsilon') \, dz \leq \int_\Omega \text{dist}(z, \Sigma_\varepsilon) \, dz - C\varepsilon^{3/2} \tag{14}$$

for some $C > 0$. Combining (13) and (14), we have

$$\int_\Omega \text{dist}(z, \Sigma_\varepsilon') \, dz \leq \int_\Omega \text{dist}(z, \Sigma_{opt}) \, dz - C\varepsilon^{3/2}$$

for some $C > 0$ when $\varepsilon \to 0^+$, which contradicts the optimality of Σ_{opt}. $\qquad \square$

Now we prove a result which in a sense is much stronger, namely, that when the length of the optimal set is sufficiently small, it must stay away from the boundary $\partial\Omega$. The result will be proven for a generic space dimension N.

Theorem 3.14. *There exist $l_0 > 0$ and $d_0 > 0$ which depend only on Ω and on N such that for all $l < l_0$ the optimal set Σ_{opt} solving the problem (7) satisfies $\text{dist}(\Sigma_{opt}, \partial\Omega) > d_0$. In particular, $\Sigma_{opt} \cap \partial\Omega = \emptyset$.*

Proof. Consider the functionals F_l defined over compact connected subsets $\Sigma \subset \bar{\Omega}$ according to the formula

$$F_l(\Sigma) := \begin{cases} \displaystyle\int_\Omega \text{dist}(x, \Sigma) \, dx, & \text{if } \mathcal{H}^1(\Sigma) \leq l, \\ +\infty, & \text{otherwise.} \end{cases}$$

As $l \to 0$, these functionals Γ^--converge to a functional

$$F_0(\Sigma) := \begin{cases} \displaystyle\int_\Omega |x - P| \, dx, & \text{if } \Sigma = \{P\} \text{ consists of one point,} \\ +\infty, & \text{otherwise.} \end{cases}$$

In fact, if $\mathcal{H}^1(\Sigma_\nu) = l_\nu \searrow 0$ and $\Sigma_\nu \to \Sigma$ in the sense of Hausdorff, then $\mathcal{H}^1(\Sigma) = 0$ according to the Golab theorem, and hence Σ consists of a single point, $\Sigma = \{P\}$, while

$$F_0(\Sigma) = \lim_\nu F_{l_\nu}(\Sigma_\nu).$$

Supposing now that the assertion to be proven is false, we would have a sequence of optimal sets $\Sigma_\nu \subset \bar{\Omega}$ such that $\mathcal{H}^1(\Sigma_\nu) \to 0$ and $\text{dist}(\Sigma_\nu, \partial\Omega) \to 0$. Up to a subsequence we may assume Σ_ν to be converging in Hausdorff sense to some compact connected set Σ consisting of a single point, i.e. $\Sigma = \{P\}$. Moreover, clearly, $P \in \partial\Omega$. The above Γ-convergence implies that P is optimal in the sense that it minimizes the distance functional $\int_\Omega |x - Q| \, dx$ among all $Q \in \bar{\Omega}$. According to Lemma 3.15 below such a point should belong to Ω, which is a contradiction. $\qquad \square$

Lemma 3.15. *Consider the optimal location problem for a single point in a convex set $\bar{\Omega} \subset \mathbf{R}^N$, i.e. a problem of finding a point $P \in \bar{\Omega}$ which provides the minimum of*

$$\inf \left\{ \int_\Omega |x - Q| \, dx \; : \; Q \in \bar{\Omega} \right\}.$$

Then $P \notin \partial\Omega$.

Proof. Let $P \in \partial\Omega$ be an arbitrary point of $\partial\Omega$. We will show that it is never optimal. In fact, without loss of generality assume that P is the origin of the coordinate system $(x_1, \ldots, x_N)$ and that the first $N-1$ coordinates $(x_1, \ldots, x_{N-1})$ are in the supporting hyperplane of Ω at P. Let the x_N axis be directed so that $x_N > 0$ for all $x \in \Omega$. Consider

$$F(x) := \int_\Omega |z - x| \, dz.$$

Then for each $i = 1, \ldots, N$ one has

$$\frac{\partial F}{\partial x_i}(0, \ldots, 0) = -\int_\Omega \frac{z_i}{|z - P|} \, dz < 0$$

showing the claim.   □

3.5. Asymptotic estimates

We claim the following result on the asymptotic behaviour of the minimum value of the functional MK as the prescribed length tends to infinity, which gives a partial answer to Problem 3.4. Denote

$$V(l) := \int_\Omega \operatorname{dist}(x, \Sigma_{opt}) \, dx$$

when $\mathcal{H}^1(\Sigma_{opt}) = l$.

Theorem 3.16. *Let $N > 1$. Then*

$$c \leq V(l) l^{1/(N-1)} \leq C$$

for some positive constants c and C which depend only on N and Ω.

To prove the lower estimate announced in the above theorem, we need the following lemma.

Lemma 3.17. *Let $N > 1$ and $Q \subset \mathbf{R}^N$ be a cube. Suppose that Q is divided by a uniform grid parallel to the edges into small subcubes with the side $\varepsilon > 0$. Let $\Sigma \subset \mathbf{R}^N$ be a Lipschitz curve of length l and k be the number of subcubes which have nonempty intersection with Σ. Then one has*

$$k \leq c_1 l/\varepsilon + c_2$$

for some positive constants c_1 and c_2 which do not depend on ε and l.

Proof. Note that to intersect all the cubes of the union of $2^N + 1$ cubes one needs a curve of length at least ε. In fact, in such a union there are two cubes, the distance between which is at least ε. Therefore, since the curve of length l connects k cubes, one has

$$l \geq [k/(2^N + 1)]\varepsilon,$$

where $[\cdot]$ stands for the integer part of the number. The above estimate shows the statement. $\qquad\square$

Proof of Theorem 3.16. The proof will be achieved in two steps.

STEP 1. Let $Q \subset \Omega$ be a cube. Divide Q by a uniform grid into small subcubes with the side $\varepsilon > 0$ (a dyadic decomposition is a particular example). Without loss of generality we may consider Σ_{opt} to be parametrized as a Lipschitz curve of length at most $2l$. Then

$$\int_\Omega \mathrm{dist}\,(x, \Sigma_{opt})\, dx \geq \int_Q \mathrm{dist}\,(x, \Sigma_{opt})\, dx.$$

The latter integral is estimated as follows. For each of the subcubes $Q_\varepsilon \subset Q$ which do not intersect Σ_{opt} one has

$$\begin{aligned}
\int_{Q_\varepsilon} \mathrm{dist}\,(x, \Sigma_{opt})\, dx &\geq \int_{\alpha Q_\varepsilon} \mathrm{dist}\,(x, \Sigma_{opt})\, dx \\
&\geq \mathcal{L}^N(\alpha Q_\varepsilon)(1-\alpha)\varepsilon \quad = \quad \mathcal{L}^N(Q_\varepsilon)\alpha^N(1-\alpha)\varepsilon
\end{aligned}$$

for all $\alpha \in [0,1]$. Hence, maximizing the last expression in α, one gets

$$\int_{Q_\varepsilon} \mathrm{dist}\,(x, \Sigma_{opt})\, dx \geq C\varepsilon^{N+1}$$

for some $C > 0$. Let k' be a number of subcubes Q_ε not intersecting Σ_{opt}. Then

$$\int_Q \mathrm{dist}\,(x, \Sigma_{opt})\, dx \geq Ck'\varepsilon^{N+1}.$$

Using the estimate from Lemma 3.17 and the fact that the total number of subcubes is $C\varepsilon^{-N}$, we arrive therefore at the estimate

$$\int_\Omega \mathrm{dist}\,(x, \Sigma_{opt})\, dx \geq \int_Q \mathrm{dist}\,(x, \Sigma_{opt})\, dx \geq c_1\varepsilon - c_2\varepsilon^N l - c_3\varepsilon^{N+1}$$

for some positive constants c_1, c_2 and c_3 independent of l and ε. Since the latter estimate is valid for all $\varepsilon > 0$, we may plug in $\varepsilon := Cl^{1/(1-N)}$ for some positive constant C, which is at the moment unknown. We see then that with the choice $0 < C < (c_1/c_2)^{1/(N-1)}$ the latter estimate becomes the desired lower bound

$$\int_\Omega \mathrm{dist}\,(x, \Sigma_{opt})\, dx \geq cl^{1/(1-N)}$$

for some $c > 0$ and for sufficiently large l.

STEP 2. The upper estimate will immediately follow from the construction of a particular Σ with $\mathcal{H}^1(\Sigma) \leq Cl$ and such that

$$\int_\Omega \text{dist}\,(x, \Sigma)\, dx \leq Cl^{1/(1-N)}, \tag{15}$$

where C denotes a positive constant different in different occasions. For this purpose consider a $(N-1)$-dimensional hyperplane π intersecting Ω by some open set T. Consider a uniform grid in T parallel to the coordinate axes in the hyperplane, with the size of a cell equal to ε. Clearly, the total length of this grid is less or equal than C/ε. Let Σ stand for the union of this grid with all the line segments perpendicular to π, passing through the nodes of T and staying in Ω. Since the total length of all such line segments is bounded from above by C/ε^{N-1}, for small ε we have $\mathcal{H}^1(\Sigma) \leq C/\varepsilon^{N-1}$. Since now by construction dist $(x, \Sigma) \leq C\varepsilon$ for all $x \in \bar{\Omega}$, where C is independent of x, then

$$\int_\Omega \text{dist}\,(x, \Sigma)\, dx \leq C\varepsilon.$$

The estimate (15) follows now by setting $l := 1/\varepsilon^{N-1}$. $\square$

Acknowledgements. The work of G. Buttazzo and E. Oudet is part of the European Research Training Network "Homogenization and Multiple Scales" (HMS2000) under contract HPRN-2000-00109. The second author has done this work while he was staying at the University of Pisa, Italy, within the framework of the above program.

Appendix A. Proof of Proposition 3.7

In order to prove Proposition 3.7, we consider the deformation of the segment S_l described in Figure 8. Namely, for all $\varepsilon > 0$, we define the polygonal line L_ε with vertices

$$(-l/2 + \delta(\varepsilon), 0), (0, \varepsilon) \text{ and } (l/2 - \delta(\varepsilon), 0), \text{ where } \delta(\varepsilon) = l/2 - \sqrt{l^2/4 - \varepsilon^2}.$$

Obviously, $\mathcal{H}^1(L_\varepsilon) = \mathcal{H}^1(S_l)$ for all $\varepsilon > 0$.

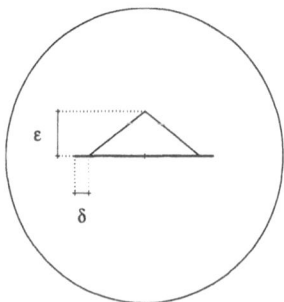

FIGURE 8. The deformation of the segment

We will now prove that for ε small enough,

$$\int_D \text{dist}(z, L_\varepsilon)dz < \int_D \text{dist}(z, S_l)dz.$$

Unfortunately, it is clear that the function $f : [-\eta, \eta] \to \mathbb{R}$ defined by

$$f(\varepsilon) := \int_D \text{dist}(z, L_\varepsilon)dz,$$

is an even function with respect to ε. So we must expect its first derivative (once it is established that f is differentiable) in $\varepsilon = 0$ to be zero. Therefore, to prove our assumption, we will need to compute the asymptotic expansion of f in 0^+ up to the order 2. We assume until the end of the proof D is the half-disc.

We consider the partition of the half-disc defined by Figure 9.

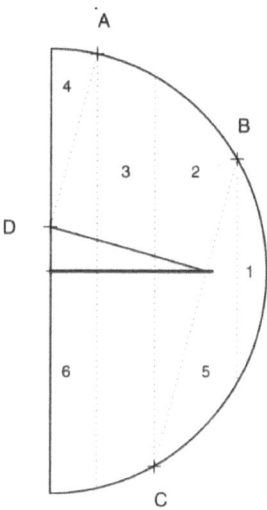

FIGURE 9. Partition of the half-disk

For each region D_i which of course depend of ε, we will compute the asymptotic expansion of $\int_{D_i(\varepsilon)} \text{dist}(z, L_\varepsilon)dz$ in ε as $\varepsilon \to 0^+$.

STEP 1. We calculate the derivatives of first order with respect to ε of the above integrals. For this purpose we first consider the following asymptotic expansion in $\varepsilon = 0^+$ (the points A, B, C, D are defined in Figure 9):

$$x_A(\varepsilon) = 2\varepsilon/l - 2\varepsilon^2/l + o(\varepsilon^2),$$
$$x_B(\varepsilon) = l/2 + 2(1 - l^2/4)^{1/2}/l\varepsilon - 3\varepsilon^2/l + o(\varepsilon^2),$$
$$x_C(\varepsilon) = l/2 - 2(1 - l^2/4)^{1/2}/l\varepsilon - 3\varepsilon^2/l + o(\varepsilon^2),$$
$$y_A(\varepsilon) = 1 - 2\varepsilon^2/l^2 + o(\varepsilon^2),$$

$$y_B\left(\varepsilon\right) = \left(1 - l^2/4\right)^{1/2} - \varepsilon + o\left(\varepsilon\right),$$

$$y_C\left(\varepsilon\right) = -\left(1 - l^2/4\right)^{1/2} - \varepsilon + o\left(\varepsilon\right).$$

Furthermore, in each of the regions D_i it is possible to evaluate explicitly the distance $d(x, y, \varepsilon)$ from the point (x, y) to the polygonal line L_ε, namely

$$d\left(x, y, \varepsilon\right) := \begin{cases} \left((x - l/2 + \delta\left(\varepsilon\right))^2 + y^2\right)^{1/2}, & (x, y) \in D_1 \cup D_5, \\ \left(x^2 + (y - \varepsilon)^2\right)^{1/2}, & (x, y) \in D_4, \\ \left(\frac{(x\varepsilon + (y - \varepsilon)(l/2 - \delta(\varepsilon)))^2}{\varepsilon^2 + (l/2 - \delta(\varepsilon))^2}\right)^{1/2}, & (x, y) \in D_2 \cup D_3 \cup D_6. \end{cases}$$

With these notations we can compute the first derivative of the following integrals

$$\frac{d}{d\varepsilon} \int_{D_1(\varepsilon)} d\left(x, y, 0\right) dx dy \bigg|_{\varepsilon = 0+} = -2\left(1 - l^2/4\right)^{3/2}/l,$$

$$\frac{d}{d\varepsilon} \int_{D_2(\varepsilon)} d\left(x, y, 0\right) dx dy \bigg|_{\varepsilon = 0+} = 2\left(1 - l^2/4\right)^{3/2}/l,$$

$$\frac{d}{d\varepsilon} \int_{D_3(\varepsilon)} d\left(x, y, 0\right) dx dy \bigg|_{\varepsilon = 0+} = -2/l - 2\left(1 - l^2/4\right)^{3/2}/l,$$

$$\frac{d}{d\varepsilon} \int_{D_4(\varepsilon)} d\left(x, y, 0\right) dx dy \bigg|_{\varepsilon = 0+} = 2/3l,$$

$$\frac{d}{d\varepsilon} \int_{D_5(\varepsilon)} d\left(x, y, 0\right) dx dy \bigg|_{\varepsilon = 0+} = 2\left(1 - l^2/4\right)^{3/2}/l,$$

$$\frac{d}{d\varepsilon} \int_{D_6(\varepsilon)} d\left(x, y, 0\right) dx dy \bigg|_{\varepsilon = 0+} = 4/3l.$$

Analogous calculus gives

$$\frac{d}{d\varepsilon} d\left(x, y, \varepsilon\right) \bigg|_{\varepsilon = 0+} = \begin{cases} 0, & (x, y) \in D_1 \cup D_5, \\ -y/\sqrt{x^2 + y^2}, & (x, y) \in D_4, \\ 2\left(x - l/2\right) \operatorname{sign} y/l, & (x, y) \in D_2 \cup D_3 \cup D_6. \end{cases}$$

Then we have

$$f\left(\varepsilon\right) = \int_D d(z, L_\varepsilon) dz = \sum_{i=1}^{6} \int_{D_i(\varepsilon)} \operatorname{dist}(z, L_\varepsilon) dz$$

$$= \sum_{i=1}^{6} \int_{D_i(0)} d(z,0) \, dz + \varepsilon \sum_{i=1}^{6} \int_{D_i(0)} \frac{d}{d\varepsilon} d(z,\varepsilon) \Big|_{\varepsilon=0+} dz$$

$$+ \varepsilon \sum_{i=1}^{6} \frac{d}{d\varepsilon} \int_{D_i(\varepsilon)} d(z,0) \, dz \Big|_{\varepsilon=0+} + o(\varepsilon) \quad = \quad \int_D \mathrm{dist}(z, S_l) dz + o(\varepsilon).$$

As expected, f is differentiable in 0^+ and its first derivative in this point is zero.

STEP 2. We have now to calculate the derivatives of second order. Estimates similar to those made in Step 1 give

$$\frac{d^2}{d\varepsilon^2} \int_{D_1(\varepsilon)} d(x,y,0) \, dxdy \Big|_{\varepsilon=0+} \quad = \quad -10 \left(l^2/4 - 1 \right) / l,$$

$$\frac{d^2}{d\varepsilon^2} \int_{D_2(\varepsilon)} d(x,y,0) \, dxdy \Big|_{\varepsilon=0+} \quad = \quad -4 \left(l^2/4 - 1 \right) / l,$$

$$\frac{d^2}{d\varepsilon^2} \int_{D_3(\varepsilon)} d(x,y,0) \, dxdy \Big|_{\varepsilon=0+} \quad = \quad 2 \left(5l^2/4 - 3 \right) / l,$$

$$\frac{d^2}{d\varepsilon^2} \int_{D_i(\varepsilon)} d(x,y,0) \, dxdy \Big|_{\varepsilon=0+} \quad = \quad -2/l, \qquad i = 4,6,$$

$$\frac{d^2}{d\varepsilon^2} \int_{D_5(\varepsilon)} d(x,y,0) \, dxdy \Big|_{\varepsilon=0+} \quad = \quad 4 \left(l^2/4 - 1 \right) / l.$$

and

$$\frac{d^2}{d\varepsilon^2} d(x,y,\varepsilon) \Big|_{\varepsilon=0+} = \begin{cases} \dfrac{x-l/2}{\left(x^2 - xl + y^2 l/2 + l^2/4\right)^{1/2}}, & (x,y) \in D_1 \cup D_5, \\ x^2 / \left(x^2 + y^2\right)^{3/2}, & (x,y) \in D_4, \\ -4\mathrm{sign}\, y/yl^2, & (x,y) \in D_2 \cup D_3 \cup D_6. \end{cases}$$

It remains to compute the following terms

$$\frac{d}{d\varepsilon} \int_{D_i(\varepsilon)} \frac{d}{d\varepsilon} d(x,y,\varepsilon) \Big|_{\varepsilon=0+} dxdy \Big|_{\varepsilon=0+} \quad = \quad 0, \ i = 1, \ 2, \ 3, \ 5,$$

$$\frac{d}{d\varepsilon} \int_{D_4(\varepsilon)} \frac{d}{d\varepsilon} d(x,y,\varepsilon) \Big|_{\varepsilon=0+} dxdy \Big|_{\varepsilon=0+} \quad = \quad -1/l,$$

$$\frac{d}{d\varepsilon} \int_{D_6(\varepsilon)} \frac{d}{d\varepsilon} d(x,y,\varepsilon) \Big|_{\varepsilon=0+} dxdy \Big|_{\varepsilon=0+} \quad = \quad 1/l.$$

Summing up, we have

$$
\begin{aligned}
f(\varepsilon) &= \int_D \operatorname{dist}(z, L_\varepsilon) dz = \sum_{i=1}^{6} \int_{D_i(\varepsilon)} \operatorname{dist}(z, L_\varepsilon) dz \\
&= \sum_{i=1}^{6} \int_{D_i(0)} d(z,0)\, dz + \varepsilon^2 \sum_{i=1}^{6} \left. \frac{d}{d\varepsilon} \int_{D_i(\varepsilon)} \left. \frac{d}{d\varepsilon} d(z,\varepsilon) \right|_{\varepsilon=0+} dz \right|_{\varepsilon=0+} \\
&\quad + \varepsilon^2 \sum_{i=1}^{6} \left. \frac{d^2}{d\varepsilon^2} \int_{D_i(\varepsilon)} d(z,0)\, dz \right|_{\varepsilon=0+} \\
&\quad + \varepsilon^2 \sum_{i=1}^{6} \int_{D_i(0)} \left. \frac{d^2}{d\varepsilon^2} d(z,\varepsilon) \right|_{\varepsilon=0+} dz + o\left(\varepsilon^2\right) \\
&= \int_D \operatorname{dist}(z, S_l) dz + \varepsilon^2 \int_{D_1(0)} \left. \frac{d^2}{d\varepsilon^2} d(z,\varepsilon) \right|_{\varepsilon=0+} dz \\
&\quad + \varepsilon^2 \int_{D_3(0)} \left. \frac{d^2}{d\varepsilon^2} d(z,\varepsilon) \right|_{\varepsilon=0+} dz + o\left(\varepsilon^2\right).
\end{aligned}
$$

STEP 3. It remains to estimate the sign of the coefficient of ε^2. That is, we have to find the sign of the quantity

$$
\begin{aligned}
\alpha_2 &:= \int_{D_1(0)} \frac{2(x - l/2)}{l\left(x^2 - xl + l^2/4 + y^2\right)^{1/2}} \, dy dx - \int_{D_3(0)} \frac{\operatorname{sign} y}{y l^2/4} \, dy dx \\
&= \int_{1/2}^{1} \int_{-\sqrt{1-x^2}}^{\sqrt{1-x^2}} \frac{2(x - l/2)}{l\left(x^2 - xl + l^2/4 + y^2\right)^{1/2}} \, dy dx \\
&\quad - \int_{0}^{l/2} \int_{-\sqrt{1-x^2}}^{\sqrt{1-x^2}} \frac{\operatorname{sign} y}{y l^2/4} \, dy dx.
\end{aligned}
\tag{16}
$$

To conclude the proof, we will show that for l small enough the above quantity is always strictly negative. We will compute its asymptotic expansion in $l = 0^+$. On one hand

$$
\begin{aligned}
&\frac{2(x - l/2)}{l\left(x^2 - xl + l^2/4 + y^2\right)^{1/2}} \\
&\quad = \frac{2x}{\left(x^2 + y^2\right)^{1/2}} l^{-1} - \frac{y^2}{\left(x^2 + y^2\right)^{3/2}} - \frac{3}{4} \frac{xy^2}{\left(x^2 + y^2\right)^{5/2}} l + o(l),
\end{aligned}
$$

and again computing derivatives we have

$$
\begin{aligned}
&\int_{1/2}^{1} \int_{-\sqrt{1-x^2}}^{\sqrt{1-x^2}} \frac{2(x - l/2)}{l\left(x^2 - xl + l^2/4 + y^2\right)^{1/2}} \, dy dx \\
&\quad = \frac{2}{l} - \int_{0}^{1} \int_{-\sqrt{1-x^2}}^{\sqrt{1-x^2}} \frac{y^2}{\left(x^2 + y^2\right)^{3/2}} \, dy dx + o(1).
\end{aligned}
$$

And on the other hand,

$$\int_0^{l/2} \int_{-\sqrt{1-x^2}}^{\sqrt{1-x^2}} \frac{\operatorname{sign} y}{yl^2/4} dy dx = \frac{2}{3} \frac{l^2/4 - 3}{l}.$$

So we have shown that

$$\alpha_2 = - \int_0^1 \int_{-\sqrt{1-x^2}}^{\sqrt{1-x^2}} \frac{y^2}{(x^2+y^2)^{3/2}} dy dx + o(1)$$

which concludes the proof.   □

Appendix B. More numerical results

Here we present some numerical results for optimal sets in a unit square of $\mathbf{R}^2$ and in a unit ball in $\mathbf{R}^3$ obtained with the use of the same evolutionary algorithms with adaptive penalty method that were employed to get optimal sets in a unit disc of $\mathbf{R}^2$. We see that these results confirm our expectations about the qualitative properties of optimal sets. In fact, the numerical approximations of optimal sets obtained are just unions of finite number of injective curves joined by triple points (i.e. points where three curves meet at an angle of 120 degrees), and they never touch the boundary of the ambient set. Moreover, it seems that if the length of the optimal set is sufficiently small, then this set contains no triple points.

FIGURE 10. Optimal sets of length 0.5 and 1 in a unit square

FIGURE 11. Optimal sets of length 1.5 and 2.5 in a unit square

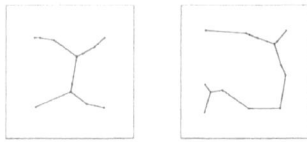

FIGURE 12. Optimal sets of length 3 and 4 in a unit square

FIGURE 13. Optim ... of length 1 and 2 in the unit ball of $\mathbf{R}^3$

FIGURE 14. Optimal sets of length 3 and 4 in the unit ball of $\mathbf{R}^3$

References

[1] L. Ambrosio, *Lecture notes on optimal transport problems*. Preprint 32, Scuola Normale Superiore, Pisa, 2000.

[2] L. Ambrosio, N. Fusco, D Pallara, *Functions of bounded variations and free discontinuity problems*. Oxford Mathematical Monographs, The Clarendon Press, Oxford University Press, New York, 2000.

[3] L. Ambrosio, P. Tilli, *Selected Topics on "Analysis in Metric Spaces"*. Scuola Normale Superiore, Pisa, Italy, 2000.

[4] G. Bouchitte, G. Buttazzo, *Characterization of optimal shapes and masses through Monge-Kantorovich equation*. J. Eur. Math. Soc., **3** (2001), 139–168.

[5] L. C. Evans, *Partial differential equations and Monge-Kantorovich mass transfer*. Current Developments in Mathematics, Cambridge MA (1997), 65–126, Int. Press, Boston MA (1999).

[6] L. C. Evans, W. Gangbo, *Differential Equations Methods for the Monge-Kantorovich Mass Transfer Problem*. Mem. Amer. Math. Soc. **137**, Providence (1999).

[7] W. Gangbo, R. J. McCann, *The geometry of optimal transportation*. Acta Math., **177** (1996), no. 2, 113–161.

[8] B. Kawohl, L. Tartar, O. Pironneau, J.-P. Zolesio, *"Optimal Shape Design"*. Springer-Verlag, Berlin, 2001.

[9] F. Morgan, R. Bolton, *Hexagonal economic regions solve the location problem*. Amer. Math. Monthly, 109:2 (2001), 165–172.

[10] E. Oudet, *Ph.D thesis on "Shape Optimization and Control"*. ULP, Strasbourg, France, 2002. In preparation.

[11] M. Schoenauer, H. Amda, *"Adaptive techniques for Evolutionary Topological Optimum Design"*. In preparation.

[12] S.T. Rachev, L. Rüschendorf, *Mass transportation problems. Vol. I Theory, Vol. II Applications*. Probability and its Applications, Springer-Verlag, Berlin (1998).

[13] J. Urbas, *Mass transfer problems*. Unpublished manuscript.

[14] C. Villani, *Topics in mass transportation*. Book in preparation.

Giusppe Buttazzo
Dipartimento di Matematica
Università di Pisa
via Buonarroti 2
I-56127 Pisa
Italy
E-mail address: buttazzo@dm.unipi.it

Edouard Oudet
Département de Mathématiques
Université Louis Pasteur
7 rue René Descartes
F-67000 Strasbourg
France
E-mail address: oudet@irma.u-strasbg.fr

Eugene Stepanov
Dipartimento di Matematica
Università di Pisa
via Buonarroti 2
I-56127 Pisa
Italy
E-mail address: stepanov@cibs.sns.it

Progress in Nonlinear Differential Equations
and Their Applications, Vol. 51, 67–80
© 2002 Birkhäuser Verlag Basel/Switzerland

Local Minimizers for a Free Gradient Discontinuity Problem in Image Segmentation

Michele Carriero, Antonio Leaci and Franco Tomarelli

Abstract. We collect the main features of the Blake & Zisserman functional and prove further necessary conditions satisfied by local minimizers.

1. Introduction

We focus the Blake & Zisserman functional in image segmentation ([3], [6]). In previous papers we proved the existence of minimizers, we showed some regularity properties ([6]–[10]) and necessary conditions for minimality ([11], [12]) by performing various kind of first variations: Euler equations and several integral and geometric conditions on optimal segmentation set.

The strong formulation of Blake & Zisserman functional F for monochromatic images and its main part E are ([8]):

$$F(K_0, K_1, u) := \int_{\Omega \setminus (K_0 \cup K_1)} \left(|D^2 u|^2 + \mu |u - g|^q \right) dx$$
$$+ \alpha \mathcal{H}^{n-1}(K_0 \cap \Omega) + \beta \mathcal{H}^{n-1}((K_1 \setminus K_0) \cap \Omega), \quad (1)$$

$$E(K_0, K_1, u) := \int_{\Omega \setminus (K_0 \cup K_1)} |D^2 u|^2 \, dx + \alpha \mathcal{H}^{n-1}(K_0 \cap \Omega) + \beta \mathcal{H}^{n-1}((K_1 \setminus K_0) \cap \Omega),$$
$$(2)$$

where $\Omega \subset \mathbf{R}^n$ is an open set, $n \geq 2$, $\mathcal{H}^{n-1}$ denotes the $(n-1)$-dimensional Hausdorff measure, and $\alpha, \beta, \mu, q \in \mathbf{R}$, with given

$$q > 1 , \ \mu > 0 , \ 0 < \beta \leq \alpha \leq 2\beta , \ g \in L^q(\Omega) , \quad (3)$$

while $K_0, \ K_1 \subset \mathbf{R}^n$ are Borel sets (a priori unknown) with $K_0 \cup K_1$ closed, $u \in C^2(\Omega \setminus (K_0 \cup K_1))$ and u is approximately continuous on $\Omega \setminus K_0$.

If (K_0, K_1, u) is a minimizing triplet for F and $n = 2, 3$, then $K_0 \cup K_1$ can be interpreted as an optimal segmentation of the monochromatic image of brightness intensity g.

Work partially supported by Italian MURST project (COFIN 2000-2001) "Calcolo delle Variazioni", by Politecnico di Milano Project "Analisi di Problemi Variazionali e Differenziali" and by GNAMPA.

Existence of minimizers for functional (1) was proved by regularization of a solution for a weak formulation, when $n = 2$, provided the additional assumption $g \in L^{2q}_{loc}(\Omega)$ is satisfied (Theorems 2.4, 2.6). In general, when g does not belong to $L^{nq}_{loc}(\Omega)$, then the infimum may be not attained ([9]).

In Section 3 we prove a Caccioppoli inequality (Theorem 3.2). Then we prove some Liouville-type properties: any local minimizer of (2) in $\mathbf{R}^n$ which has finite energy and bounded singular sets is affine (Theorem 3.3); any local minimizer of (2) in $\mathbf{R}^2$ which is bi-harmonic is affine too (Theorem 3.5). Moreover we prove that the step and the wedge are not local minimizers of the main part of the functional in $\mathbf{R}^n$. Eventually we show a candidate local minimizer of E compatible with all the necessary conditions (see [11]).

In Section 4 we show statements for the vector-valued case (Theorems 4.1, 4.2) which are relevant for color images.

In Section 5 we improve the sufficient conditions for the existence of strong minimizers of functionals (1) and (17) (Theorems 5.1, 5.2).

2. Notation, definitions and recent results

We denote by $B_\rho(\mathbf{x}_0)$ the open ball with radius $\rho > 0$ centered at $\mathbf{x}_0$.

For any Borel function $v : \Omega \to \mathbf{R}$ and $\mathbf{x} \in \Omega$, $z \in \overline{\mathbf{R}} := \mathbf{R} \cup \{-\infty, +\infty\}$, we set $z = \operatorname{ap\,lim}_{\mathbf{y} \to \mathbf{x}} v(\mathbf{y})$ (the approximate limit of v at $\mathbf{x}$ denoted also by $\tilde{v}(\mathbf{x})$) if

$$\psi(z) = \lim_{\rho \to 0} \fint_{B_\rho(0)} \psi(v(\mathbf{x} + \mathbf{y})) d\mathbf{y} \qquad \forall \psi \in C^0(\overline{\mathbf{R}}).$$

The set $S_v = \{\mathbf{x} \in \Omega : \operatorname{ap\,lim}_{\mathbf{y} \to \mathbf{x}} v(\mathbf{y}) \text{ does not exist}\}$ is the singular set of v. By $Dv, \nabla v$ we denote, respectively, the distributional gradient and the approximate gradient of v (see [8]). $|\cdot|$ denotes the euclidean norm and $\nabla_i v = (\mathbf{e}_i \cdot \nabla)v$, where $\{\mathbf{e}_i\}$ is the canonical basis of $\mathbf{R}^n$. When the right-hand side is meaningful, we set $\nabla^2_{ij} v = \nabla_i(\nabla_j v)$. We recall also the definitions of some classes of functions having derivatives which are special measures in the sense of De Giorgi, and we refer to [2], [4], [6]–[9], [13], [14], for their properties:

$$SBV(\Omega) := \left\{ v \in BV(\Omega) : \|Dv\|_{\mathcal{M}(\Omega)} = \int_\Omega |\nabla v| \, d\mathbf{x} + \int_{S_v} |v^+ - v^-| \, d\mathcal{H}^{n-1} \right\},$$

$$GSBV(\Omega) := \left\{ v : \Omega \to \mathbf{R} \text{ Borel function}; -k \vee v \wedge k \in SBV_{loc}(\Omega) \ \forall k \in \mathbf{N} \right\},$$

$$GSBV^2(\Omega) := \left\{ v \in GSBV(\Omega); \ \nabla v \in \left(GSBV(\Omega)\right)^n \right\}.$$

The classes $GSBV(\Omega)$, $GSBV^2(\Omega)$ are neither vector spaces nor subsets of distributions in Ω, nevertheless smooth variations of a function in $GSBV^2(\Omega)$ belong to the same class. If $v \in GSBV(\Omega)$, then S_v is countably $\mathcal{H}^{n-1}$-rectifiable and ∇v exists a.e. in Ω. We set $S_{\nabla v} = \bigcup_{i=1}^n S_{\nabla_i v}$, and $K_v = \overline{S_v \cup S_{\nabla v}}$.

Definition 2.1 (Weak formulation of Blake & Zisserman functional [7]). *For $\Omega \subset \mathbf{R}^n$ open set, under the assumption (3), we define $\mathcal{F} : X(\Omega) \to [0, +\infty]$ by*

$$\mathcal{F}(v) := \int_\Omega (|\nabla^2 v|^2 + \mu|v - g|^q)\, d\mathbf{x} + \alpha \mathcal{H}^{n-1}(S_v) + \beta \mathcal{H}^{n-1}(S_{\nabla v} \setminus S_v) \qquad (4)$$

where $X(\Omega) := GSBV^2(\Omega) \cap L^q(\Omega)$.

We consider also the localization $\mathcal{F}_A$ of $\mathcal{F}$ on any Borel set $A \subseteq \Omega$:

$$\mathcal{F}_A(v) := \int_A (|\nabla^2 v|^2 + \mu|v - g|^q)\, d\mathbf{x} + \alpha \mathcal{H}^{n-1}(S_v \cap A) + \beta \mathcal{H}^{n-1}((S_{\nabla v} \setminus S_v) \cap A).$$

We remark that the subset of $GSBV^2(\Omega)$ where $\mathcal{F}$ is finite is a vector space, while $GSBV^2(\Omega)$ is not a vector space.

Definition 2.2 (Local minimizer). *We say that u is a local minimizer of the functional $\mathcal{F}$ in Ω if*

$$u \in GSBV^2(A), \qquad \mathcal{F}_A(u) < +\infty, \qquad \mathcal{F}_A(u) \leq \mathcal{F}_A(u + \varphi) \qquad (5)$$

for every open subset $A \subset\subset \Omega$ and for every $\varphi \in GSBV^2(\Omega)$ with compact support in A.

We introduce also the weak form of functional (2)

$$\mathcal{E}(v) := \int_\Omega |\nabla^2 v|^2\, d\mathbf{x} + \alpha \mathcal{H}^{n-1}(S_v) + \beta \mathcal{H}^{n-1}(S_{\nabla v} \setminus S_v). \qquad (6)$$

We say that u is a local minimizer of the functional $\mathcal{E}$ in Ω if, by denoting $\mathcal{E}_A$ the localization of $\mathcal{E}$,

$$u \in GSBV^2(\Lambda), \qquad \mathcal{E}_A(u) < +\infty, \qquad \mathcal{E}_A(u) \leq \mathcal{E}_A(u + \varphi) \qquad (7)$$

for every open subset $A \subset\subset \Omega$ and for every $\varphi \in GSBV^2(\Omega)$ with compact support in A.

Remark 2.3. *If u is a local minimizer of $\mathcal{E}$ in Ω then also the function $u(\mathbf{x}) + a + \mathbf{b} \cdot \mathbf{x}$ is a local minimizer in Ω for every $a \in \mathbf{R}, \mathbf{b} \in \mathbf{R}^n$; moreover, if $B_\rho(\mathbf{x}_0) \subset \Omega$, then the re-scaling*

$$u_\rho(\mathbf{x}) = \rho^{-3/2} u(\mathbf{x}_0 + \rho \mathbf{x})$$

defines a local minimizer of $\mathcal{E}$ in $B_1(0)$ such that $\mathcal{E}_{B_\rho(\mathbf{x}_0)}(u) = \rho^{n-1} \mathcal{E}_{B_1(0)}(u_\rho)$.

We recall some statements about the minimization of (4) and (1).

Theorem 2.4 (Existence of weak solutions [7]). *Let $\Omega \subset \mathbf{R}^n$ be an open set and assume (3). Then there is $v_0 \in X(\Omega)$ such that $\mathcal{F}(v_0) \leq \mathcal{F}(v) \ \forall v \in X(\Omega)$.*

We recall that assumption $\beta \leq \alpha \leq 2\beta$ is necessary for lower semi-continuity of $\mathcal{F}$.

Remark 2.5. *Theorem 2.4 holds true even if $|\nabla^2 v|^2$ is substituted by a positive definite quadratic form evaluated in $\nabla^2 v$.*

Theorem 2.6 (Existence of strong solutions [8]). *Let $n = 2$, $\Omega \subset \mathbf{R}^2$ be an open set. Assume (3) and $g \in L^{2q}_{loc}(\Omega)$. Then there is at least one triplet among $K_0, K_1 \subset \mathbf{R}^2$ Borel sets with $K_0 \cup K_1$ closed, u approximately continuous on $\Omega \setminus K_0$ and $u \in C^2(\Omega \setminus (K_0 \cup K_1))$ minimizing the functional (1) with finite energy. Moreover the sets $K_0 \cap \Omega$ and $K_1 \cap \Omega$ are $(\mathcal{H}^1, 1)$ rectifiable.*

In the following theorems we state some necessary conditions satisfied by local minimizers of $\mathcal{F}$.

Theorem 2.7 (Euler equation and regularity outside the optimal segmentation K_u [11], [12]). *Assume (3) and $u \in GSBV^2(\Omega)$ is a local minimizer of $\mathcal{F}$ in $\Omega \subset \mathbf{R}^n$, $n \geq 2$, then*

(i) $\Delta^2 u = -\frac{q}{2} \mu |u - g|^{q-2}(u - g)$ *in $\Omega \setminus K_u$;*

(ii) *there exists a constant $c > 0$ (depending on n, α, β, μ, g) such that for every open ball $B \subset \Omega \setminus K_u$*

$$\sup_{\substack{x,y \in B \\ x \neq y}} \frac{|Du(x) - Du(y)|}{|x - y|^{1/2}} \leq c,$$

hence $u \in C^{1,1/2}_{loc}(\Omega \setminus K_u)$;

(iii) $u \in W^{4,q/(q-1)}_{loc}(\Omega \setminus K_u)$.

Moreover if $s \geq n(q - 1)$ and $g \in L^s(\Omega)$, then, by setting $\gamma = 1 - \frac{n(q-1)}{s}$,

(iv) $u \in W^{4,s/(q-1)}_{loc}(\Omega \setminus K_u) \subset C^{3,\gamma}_{loc}(\Omega \setminus K_u)$.

If A is a C^2 uniformly regular open subset of Ω , N is the outward unit normal to ∂A and $\{t_k = t_k(x) \; ; \; k = 1, \ldots, n - 1, x \in \partial A \}$ denotes a system of local tangential co-ordinates, then for every $\varphi \in W^{2,2}(A)$ and $u \in W^{2,2}(A)$ with $\Delta^2 u \in L^2(A)$ the following **Green formula** holds true:

$$\int_A (D^2 u) : (D^2 \varphi) \, dy = \int_A (\Delta^2 u)\varphi \, dy$$

$$+ \int_{\partial A} \left(S(u) - \frac{\partial}{\partial N}\Delta u\right)\varphi \, d\mathcal{H}^{n-1} + \int_{\partial A} T(u)\frac{\partial \varphi}{\partial N} \, d\mathcal{H}^{n-1}$$

where the natural boundary operators $T(u)$ and $S(u)$ are defined by

$$T(u) := \sum_{i,j=1}^{n} \nabla^2_{ij} u \, N_i N_j \, , \quad S(u) := -\sum_{i,j=1}^{n} \sum_{k=1}^{n-1} \frac{\partial}{\partial t_k}\left(\nabla^2_{ij} u \, N_j \frac{\partial t_k}{\partial x_i}\right) . \qquad (8)$$

By evaluating the first variation of the energy functional (4) around a local minimizer u of $\mathcal{F}$ (Definition 2.2) under deformations of u, which have compact support and are smooth outside K_u , we proved the following theorems.

Theorem 2.8 (Necessary conditions on S_u for natural boundary operators [11], [12]). *Assume (3), $n \geq 2$, $q > 1$ and u is a local minimizer of $\mathcal{F}$, $B \subset\subset \Omega$ an open ball such that $S_u \cap B$ is the graph of a C^3 function and $(S_{\nabla u} \setminus S_u) \cap B = \emptyset$. Denote by B^+, B^- the two connected components of $B \setminus S_u$ and by N the unit normal vector to S_u pointing toward B^+. Assume that $u \in C^3(\overline{B^+}) \cap C^3(\overline{B^-})$. Then*

$$\left(T(u)\right)^{\pm} = 0\,, \qquad \left(S(u) - \frac{\partial}{\partial N}\Delta u\right)^{\pm} = 0 \qquad on\ S_u \cap B\,. \tag{9}$$

Theorem 2.9 (Necessary conditions on $S_{\nabla u}$ for jumps of natural boundary operators [11], [12]). *Assume (3), $n \geq 2$, $q > 1$ and u is a local minimizer of $\mathcal{F}$, $B \subset\subset \Omega$ an open ball such that $S_{\nabla u} \cap B$ is the graph of a C^3 function and $S_u \cap B = \emptyset$. Denote by B^+, B^- the two connected components of $B \setminus S_{\nabla u}$ and by N the unit normal vector to $S_{\nabla u}$ pointing toward B^+. Assume that $u \in C^3(\overline{B^+}) \cap C^3(\overline{B^-})$. Then*

$$\left(T(u)\right)^{\pm} = 0\,, \qquad \left[\!\left[S(u) - \frac{\partial}{\partial N}\Delta u\right]\!\right] = 0 \qquad on\ S_{\nabla u} \cap B\,, \tag{10}$$

where for a function w we set $[\![w]\!] = w^+ - w^-$.

We emphasize that Theorems 2.6–2.9 hold true for local minimizers of $\mathcal{E}$ provided all the terms including $u - g$ are dropped.

3. Caccioppoli type inequality for minimizers of $\mathcal{E}$ and consequences

So far we have found many necessary conditions for local minimizers of the functionals $\mathcal{F}$ and $\mathcal{E}$. Now we focus the main part $\mathcal{E}$ of the functional $\mathcal{F}$ in $\mathbf{R}^n$, which is a natural procedure in the study of regularity properties of $\mathcal{F}$.

In this section we recall an easy estimate for the energy of a local minimizer of $\mathcal{E}$ and we prove a Caccioppoli type inequality for a local minimizer of the main part of the Blake & Zisserman functional similar to the L^2 estimate of Hessian in case of the elastic-plastic thin plate (see [5]); such inequality entails that local minimizers in $\mathbf{R}^n$, with finite energy and bounded singular sets, are affine. Moreover we show a Liouville type property for bi-harmonic functions which are local minimizers in $\mathbf{R}^2$. Eventually we prove that a 1-dimensional step or an infinite dihedral are not local minimizers of $\mathcal{E}$ in $\mathbf{R}^n$.

We denote by ω_n the volume of the unit ball in $\mathbf{R}^n$, so that $n\omega_n$ is the area of its boundary.

Theorem 3.1. *Let $\Omega \subset \mathbf{R}^n$ be an open set and let u be a local minimizer of $\mathcal{E}$ in Ω. Then for every ball $B_R \subset \Omega$ we have*

$$\mathcal{E}_{B_R}(u) \leq \alpha n \omega_n R^{n-1}.$$

Proof. By comparison with the function $v = u\chi_{\Omega \setminus B_R}$. $\qquad\square$

Theorem 3.2 (Caccioppoli inequality). *Assume $\Omega \subset \mathbf{R}^n$ and u is a local minimizer of $\mathcal{E}$ in Ω. Then, for every $a \in \mathbf{R}$, $\mathbf{b} \in \mathbf{R}^n$ and for every $\rho > 0$ such that $B_{2\rho} \subset \Omega$, we have*

$$\int_{B_\rho} |\nabla^2 u|^2 \, d\mathbf{x} \le \frac{c}{\rho^2} \int_{B_{2\rho} \setminus B_\rho} |\nabla u - \mathbf{b}|^2 \, d\mathbf{x} + \frac{c}{\rho^4} \int_{B_{2\rho} \setminus B_\rho} |u - a - \mathbf{b} \cdot \mathbf{x}|^2 \, d\mathbf{x} \, ,$$

where c is a constant independent of u and ρ.

Proof. By Remark 2.3 we can assume $a = 0$, $\mathbf{b} = \mathbf{0}$. Let $\varphi \in C_0^\infty(B_{2\rho})$ such that

$$0 \le \varphi \le 1, \quad \varphi \equiv 1 \text{ in } B_\rho, \quad |D\varphi| \le \frac{c_1}{\rho}, \quad |D^2\varphi| \le \frac{c_1}{\rho^2}.$$

For $|\varepsilon| < 1$ set $u_\varepsilon = u + \varepsilon \varphi^4 u$, then

$$\nabla u_\varepsilon = (1 + \varepsilon \varphi^4) \nabla u + 4\varepsilon u \varphi^3 D\varphi,$$
$$S_{u_\varepsilon} = S_u, \qquad S_{\nabla u_\varepsilon} = S_{\nabla u}.$$

Now we have

$$|\nabla^2 u_\varepsilon|^2 \;=\; |\nabla^2 u|^2 + 2\varepsilon \left(\varphi^4 |\nabla^2 u|^2 + 4\varphi^3 (\nabla u \nabla^2 u D\varphi + D\varphi \nabla^2 u \nabla u) \right.$$

$$\left. + 12\varphi^2 u D\varphi \nabla^2 u D\varphi + 4\varphi^3 u \nabla^2 u : D^2\varphi \right) + o(\varepsilon).$$

By the minimality of u we get

$$\int_{B_{2\rho}} \left(\varphi^4 |\nabla^2 u|^2 + 4\varphi^3 (\nabla u \nabla^2 u D\varphi + D\varphi \nabla^2 u \nabla u) \right.$$

$$\left. + 12\varphi^2 u D\varphi \nabla^2 u D\varphi + 4\varphi^3 u \nabla^2 u : D^2\varphi \right) \, d\mathbf{x} = 0.$$

By using the above equation and Hölder inequality we obtain

$$\int_{B_{2\rho}} \varphi^4 |\nabla^2 u|^2 \, d\mathbf{x}$$

$$\le c_2 \int_{B_{2\rho}} \left(|\varphi^2 \nabla^2 u| |\varphi \nabla u| |D\varphi| + |\varphi^2 \nabla^2 u| |u| |D\varphi|^2 + |\varphi^2 \nabla^2 u| |\varphi u D^2\varphi| \right) \, d\mathbf{x}$$

$$\le c_3 \left(\int_{B_{2\rho}} \varphi^4 |\nabla^2 u|^2 \, d\mathbf{x} \right)^{\frac{1}{2}}$$

$$\times \left\{ \frac{1}{\rho} \left(\int_{B_{2\rho} \setminus B_\rho} |\nabla u|^2 \, d\mathbf{x} \right)^{\frac{1}{2}} + \frac{1}{\rho^2} \left(\int_{B_{2\rho} \setminus B_\rho} u^2 \, d\mathbf{x} \right)^{\frac{1}{2}} \right\},$$

hence by Young inequality we have

$$\int_{B_\rho} |\nabla^2 u|^2 \, d\mathbf{x} \le \int_{B_{2\rho}} \varphi^4 |\nabla^2 u|^2 \, d\mathbf{x} \le \frac{c}{\rho^2} \int_{B_{2\rho} \setminus B_\rho} |\nabla u|^2 \, d\mathbf{x} + \frac{c}{\rho^4} \int_{B_{2\rho} \setminus B_\rho} u^2 \, d\mathbf{x},$$

and the inequality is proved. $\square$

By using the "hole-filling" technique, we derive a consequence of the above inequality.

Theorem 3.3. *Let u be a local minimizer of $\mathcal{E}$ in $\mathbf{R}^n$ such that S_u and $S_{\nabla u}$ are bounded and*

$$\int_{\mathbf{R}^n} |\nabla^2 u|^2 \, dx < +\infty. \tag{11}$$

Then u is affine.

Proof. Assume $S_u \cup S_{\nabla u} \subset B_{\rho_0}$ and $\rho > \rho_0$. By using Theorem 3.2 with

$$a = |B_{2\rho} \setminus B_\rho|^{-1} \int_{B_{2\rho} \setminus B_\rho} u \, dx, \qquad b = |B_{2\rho} \setminus B_\rho|^{-1} \int_{B_{2\rho} \setminus B_\rho} \nabla u \, dx$$

and applying the Poincaré inequality to $u \in W^{2,2}(B_{2\rho} \setminus B_\rho)$, we get

$$\int_{B_\rho} |\nabla^2 u|^2 \, dx \le \frac{c}{\rho^2} \int_{B_{2\rho} \setminus B_\rho} |\nabla u - b|^2 \, dx + \frac{c}{\rho^4} \int_{B_{2\rho} \setminus B_\rho} |u - a - b \cdot x|^2 \, dx$$

$$\le c' \int_{B_{2\rho} \setminus B_\rho} |\nabla^2 u|^2 \, dx.$$

Filling the hole we obtain

$$(1 + c') \int_{B_\rho} |\nabla^2 u|^2 \, dx \le c' \int_{B_{2\rho}} |\nabla^2 u|^2 \, dx,$$

so that

$$\int_{B_\rho} |\nabla^2 u|^2 \, dx \le \theta \int_{B_{2\rho}} |\nabla^2 u|^2 \, dx \quad \text{with } \theta = \frac{c'}{1 + c'} < 1,$$

and, for every $k \in \mathbf{N}$,

$$\int_{B_\rho} |\nabla^2 u|^2 \, dx \le \theta^k \int_{B_{2^k \rho}} |\nabla^2 u|^2 \, dx.$$

By the assumption (11) and the arbitrariness of k we conclude that

$$\int_{B_\rho} |\nabla^2 u|^2 \, dx = 0.$$

By the arbitrariness of ρ, the function u is affine in $\mathbf{R}^n \setminus B_{\rho_0}$. Then, by the minimality, u is affine in $\mathbf{R}^n$. $\qquad\square$

Remark 3.4. *Theorems 3.1–3.3 hold true even if $|\nabla^2 v|^2$ is substituted by a positive definite quadratic form evaluated in $\nabla^2 v$.*

Now we show a Liouville type property for local minimizers of $\mathcal{E}$ in $\mathbf{R}^2$.

Theorem 3.5. *(Liouville property) Bi-harmonic functions are local minimizers for $\mathcal{E}$ in $\mathbf{R}^2$ if and only if they are affine.*

Proof. If u is affine then $\mathcal{E}(u) = 0$, hence u is a local minimizer of $\mathcal{E}$ in $\mathbf{R}^2$. If u is bi-harmonic and local minimizer of $\mathcal{E}$ in $\mathbf{R}^2$, then by the Almansi decomposition of a bi-harmonic function (see [1]) we have, in polar co-ordinates,

$$u = v + \rho^2 w,$$

with v and w harmonic functions, so that

$$u = \sum_{k=0}^{\infty} \varrho^k \left((a_k + \alpha_k \varrho^2) \cos k\theta + (b_k + \beta_k \varrho^2) \sin k\theta \right)$$

and, by exploiting the orthogonality in $L^2(0, 2\pi)$ of the trigonometric system, for any $R > 0$ we get

$$\int_{B_R} |D^2 u|^2 \, dx \, dy$$

$$= 8\alpha_0^2 \pi R^2 + 2\pi \sum_{k=1}^{\infty} R^{2(k-1)} \Big[(k-1)k^2(a_k^2 + b_k^2)$$

$$+ 2k(k^2 - 1)R^2(a_k \alpha_k + b_k \beta_k) + (k+1)(k^2 + 2)R^4(\alpha_k^2 + \beta_k^2) \Big]. \qquad (12)$$

Since the following inequalities, and analogous ones for the terms containing $b_k \beta_k$, hold

$$|2k(k^2 - 1)R^2 a_k \alpha_k| = |2k\sqrt{k-1} a_k(k+1)\sqrt{k-1}R^2\alpha_k|$$

$$\le k^2(k-1)a_k^2 + (k+1)^2(k-1)R^4\alpha_k^2, \qquad (13)$$

by Theorem 3.1 we get

$$8\alpha_0^2 \pi R^2 + 2\pi \sum_{k=1}^{\infty} R^{2(k-1)} \Big[3(k+1)R^4(\alpha_k^2 + \beta_k^2) \Big]$$

$$\le \int_{B_R} |D^2 u|^2 \, dx \, dy = \mathcal{E}_{B_R}(u) \le 2\pi\alpha R, \qquad (14)$$

hence, by the arbitrariness of R, $\alpha_k = \beta_k = 0$ for every k. Coming back to equality (12) it reduces to

$$2\pi \sum_{k=1}^{\infty} R^{2(k-1)} \Big[(k-1)k^2(a_k^2 + b_k^2) \Big] = \int_{B_R} |D^2 u|^2 \, dx \, dy = \mathcal{E}_{B_R}(u) \le 2\pi\alpha R,$$

hence $a_k = b_k = 0$ for every $k > 1$ and in cartesian co-ordinates (x, y) we have

$$u = a_0 + a_1 x + b_1 y.$$

$\square$

Theorem 3.6. *Set* $\mathbf{x} = (x_1, \mathbf{x}')$ *for* $\mathbf{x} \in \mathbf{R}^n$. *Then, for any* $c \ne 0$, *the function* $u(\mathbf{x}) = c \, \mathrm{sign}(x_1)$ *is not a local minimizer for* $\mathcal{E}$ *in* $\mathbf{R}^n$.

Proof. If $n = 1$ we define, for any $\delta > 0$,

$$
u_1(x) = \begin{cases} c\left(1 - \dfrac{(x-\delta)^2}{\delta^2}\right) & \text{if } 0 \le x \le \delta \\[3mm] -c\left(1 - \dfrac{(x+\delta)^2}{\delta^2}\right) & \text{if } -\delta \le x < 0, \\[3mm] u(x) & \text{if } |x| > \delta. \end{cases}
$$

Then $|u_1''(x)| = 2|c|/\delta^2$ if $|x| < \delta$, hence for any $R > \delta$ we have

$$
\mathcal{E}_{BR}(u_1) - \mathcal{E}_{BR}(u) = 8\frac{c^2}{\delta^3} - \alpha < 0 \qquad \text{if} \quad \delta^3 > 8\frac{c^2}{\alpha},
$$

which contradicts the minimality of u.

If $n > 1$ we define, for any $\delta > 0$

$$
\bar{u}(\mathbf{x}) = \begin{cases} u_1(x_1) & \text{if } |x_1| \le \delta, \ |\mathbf{x}'| < L \\[2mm] u(\mathbf{x}) & \text{elsewhere .} \end{cases}
$$

Then $|\nabla^2 \bar{u}(\mathbf{x})| = |u_1''(x_1)| = 2|c|/\delta^2$ if $|x_1| < \delta$, $|\mathbf{x}'| < L$. Hence for any R with $R^2 > L^2 + \delta^2$ we have

$$
\mathcal{E}_{BR}(\bar{u}) - \mathcal{E}_{BR}(u) = \frac{8c^2}{\delta^3}\omega_{n-1}L^{n-1} + 2\alpha(n-1)\omega_{n-1}L^{n-2}\delta - \alpha\omega_{n-1}L^{n-1} < 0
$$

if we choose

$$
\delta^3 > 16\frac{c^2}{\alpha} \qquad \text{and} \qquad L > 4(n-1)\delta,
$$

and this contradicts the minimality of u. $\qquad\square$

Theorem 3.7. *Set* $\mathbf{x} = (x_1, \mathbf{x}')$ *for* $\mathbf{x} \in \mathbf{R}^n$. *Then, for any* $c \neq 0$, *the function* $d(\mathbf{x}) = c|x_1|$ *is not a local minimizer for* $\mathcal{E}$ *in* $\mathbf{R}^n$.

Proof. If $n = 1$ we define, for any $\delta > 0$,

$$
d_1(x) = \begin{cases} c\left(\dfrac{x^2}{2\delta} + \dfrac{\delta}{2}\right) & \text{if } |x| \le \delta \\[3mm] d(x) & \text{if } |x| > \delta. \end{cases}
$$

Then $|d_1''(x)| = |c|/\delta$ if $|x| < \delta$, hence for any $R > \delta$

$$
\mathcal{E}_{BR}(d_1) - \mathcal{E}_{BR}(d) = 2\frac{c^2}{\delta} - \beta < 0 \qquad \text{if } \delta > 2\frac{c^2}{\beta},
$$

which contradicts the minimality of d.

If $n > 1$ we define, for any $\delta > 0$,

$$
\bar{d}(\mathbf{x}) = \begin{cases} d_1(x_1) & \text{if } |x_1| \le \delta, \ |\mathbf{x}'| \le L \\[2mm] d(\mathbf{x}) & \text{elsewhere.} \end{cases}
$$

Then $|\nabla^2 \bar{d}(\mathbf{x})| = |d_1''(x_1)| = |c|/\delta$ if $|x_1| < \delta$, $|\mathbf{x}'| < L$. Hence, for any R with $R^2 > L^2 + \delta^2$ we have

$$\mathcal{E}_{B_R}(\bar{d}) - \mathcal{E}_{B_R}(d) =$$

$$\frac{2c^2}{\delta}\omega_{n-1}L^{n-1} + 2\alpha\delta(n-1)\omega_{n-1}L^{n-2} - \beta\omega_{n-1}L^{n-1} < 0$$

if we choose

$$\delta > 4\frac{c^2}{\beta}, \qquad L > \frac{4(n-1)\delta\alpha}{\beta}, \qquad R^2 > L^2 + \delta^2,$$

and this contradicts the minimality of d. $\qquad\square$

Theorems 2.7–2.9, and the Euler equations at crack-tip and crease-tip (see [11], [12]) impose strong geometric constraint on local minimizers of $\mathcal{F}$ and $\mathcal{E}$. Nevertheless in the analysis of the main part $\mathcal{E}$ of the functional $\mathcal{F}$ in $\mathbf{R}^2$ we showed explicitly a candidate local minimizer W of $\mathcal{E}$ in $\mathbf{R}^2$, which has a non empty jump set (see [11]). It is constructed by suitably adding real parts of analytic branch of multi-valued functions with branching point at the origin and a cut along the negative real axis, and by exploiting the Almansi representation of bi-harmonic functions (see [1]).

Such candidate, expressed by polar co-ordinates in $\mathbf{R}^2$ with $\theta \in (-\pi, \pi)$, is:

$$W = \sqrt{\frac{\alpha\rho^3}{193\pi}} \left(\sqrt{21}\left(\sin\frac{\theta}{2} - \frac{5}{3}\sin\left(\frac{3}{2}\theta\right)\right) + \left(\cos\frac{\theta}{2} - \frac{7}{3}\cos\left(\frac{3}{2}\theta\right)\right)\right). \quad (15)$$

Notice that W is left unchanged by natural dilations of homogeneity $-3/2$ (see Remark 2.3). In addition a variational principle of equi-partition of bulk and surface energy is fulfilled by W, say, for any $\varrho > 0$,

$$\int_{B_\varrho(0)} |\nabla^2 W|^2 \, dx \, dy = \alpha\mathcal{H}^1(S_W \cap B_\varrho(0)).$$

This function W exhibits the only homogeneity in ρ compatible with minimality and fulfills the following requirements: being bi-harmonic outside the singular set, scaling invariance of the energy, all the necessary conditions on the jump set, local finiteness of energy and the proper decay rate of energy around the origin (tip of the crack).

The following list of properties shows that W fulfills the necessary conditions of Theorems 2.7, 2.8, Euler condition at crack-tip and the equi-partition of energy:

$$\begin{cases}
S_W = \text{negative real axis}, \quad S_{\nabla W} = \emptyset, \\
\Delta^2 W = 0 \text{ on } \mathbf{R}^2 \setminus \overline{S_W}, \\
W_{yy} = 0, \quad W_{yyy} + 2W_{xxy} = 0, \quad \text{on both sides of } S_W, \\
W^\pm = \pm\frac{8}{3}\sqrt{\frac{21\alpha\rho^3}{193\pi}} \text{ on } S_W, \\
\int_{B_\rho(0)} |\nabla^2 W|^2 dx dy = \alpha\rho = \alpha\mathcal{H}^1(S_W \cap B_\rho(0)).
\end{cases}$$

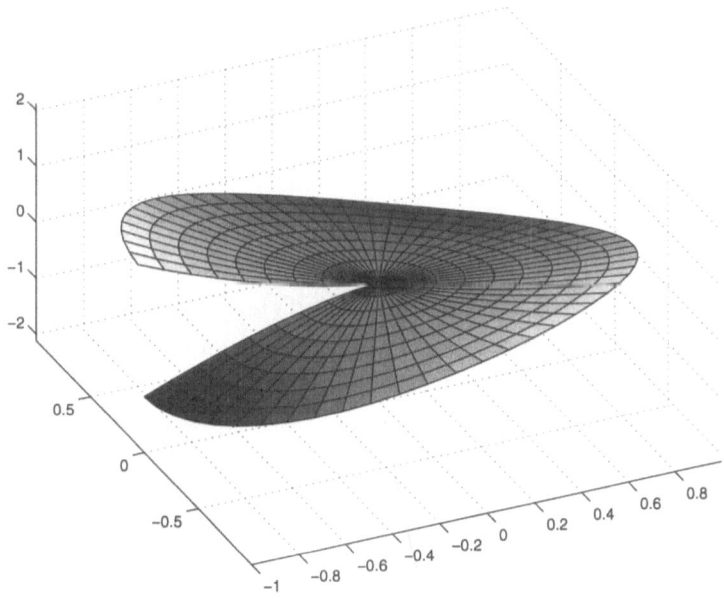

FIGURE 1. Candidate local minimizer in $\mathbf{R}^2$ (picture gray levels correspond to candidate gray levels).

We stress the fact that minimizers are not defined up to a free constant multiplier, due to the analysis around the crack-tip.

By the previous analysis and theorems 3.5, 3.6 and 3.7 we are led to the following statement.

We conjecture that W is a local minimizer of $\mathcal{E}$ in $\mathbf{R}^2$, and there are no other nontrivial local minimizers besides W, up to sign change, rigid motions in $\mathbf{R}^2$ and addition of affine functions.

All the statements obtained in this paragraph have a natural extension to local minimizing triplets of functional E defined by (2).

4. Vector-valued case: RGB color images

When processing a color image the datum is given by a vector-valued input image $\mathbf{g} : \Omega \to \mathbf{R}^3$ and one looks for a vector-valued segmented image $\mathbf{u} : \Omega \to \mathbf{R}^3$. Here the approximate limit of vector-valued functions is defined component-wise, and the singular set is the union of the singular sets of components. Assume

$$q > 1 \, , \, \mu_j > 0 \, , \, 0 < \beta_j \leq \alpha_j \leq 2\beta_j \, , \text{ for } j = 1, 2, 3 \, , \, \mathbf{g} \in \left(L^q(\Omega) \right)^3 . \qquad (16)$$

We define the strong formulation of Blake & Zisserman functional for vector-valued images as follows

$$\mathbf{F}(\mathbf{K}_0, \mathbf{K}_1, \mathbf{u}) := \sum_{j=1}^{3} \int_{\Omega \setminus (K_0^j \cup K_1^j)} \left(|D^2 u_j|^2 + \mu_j |u_j - g_j|^q \right) dx$$

$$+ \sum_{j=1}^{3} \left(\alpha_j \mathcal{H}^{n-1}(K_0^j \cap \Omega) + \beta_j \mathcal{H}^{n-1}((K_1^j \setminus K_0^j) \cap \Omega) \right), \quad (17)$$

where $\mathbf{K}_0 = (K_0^1, K_0^2, K_0^3)$, $\mathbf{K}_1 = (K_1^1, K_1^2, K_1^3)$, $\mathbf{u} = (u_1, u_2, u_3)$ with the properties K_0^j, $K_1^j \subset \mathbf{R}^n$ Borel sets, $K_0^j \cup K_1^j$ closed, u_j approximately continuous on $\Omega \setminus K_0^j$ and $u_j \in C^2(\Omega \setminus (K_0^j \cup K_1^j))$ $(j = 1, 2, 3)$.

We define the weak formulation of Blake & Zisserman functional for vector-valued images as follows

$$\mathcal{F}(\mathbf{v}) := \sum_{j=1}^{3} \int_{\Omega} \left(|\nabla^2 v_j|^2 + \mu_j |v_j - g_j|^q \right) dx$$

$$+ \sum_{j=1}^{3} \left(\alpha_j \mathcal{H}^{n-1}(S_{v_j}) + \beta_j \mathcal{H}^{n-1}(S_{\nabla v_j} \setminus S_{v_j}) \right), \quad (18)$$

where $\mathbf{v} \in \mathbf{X}(\Omega) := \left(GSBV^2(\Omega) \right)^3 \cap \left(L^q(\Omega) \right)^3$.

By the same techniques of [7] and [8] we can prove the following statements.

Theorem 4.1 (Existence of weak segmentation for RGB color images). *Let $\Omega \subset \mathbf{R}^n$ be an open set and assume (16). Then there is $\mathbf{w} \in \mathbf{X}(\Omega)$ such that $\mathcal{F}(\mathbf{w}) \leq \mathcal{F}(\mathbf{v})$ for any $\mathbf{v} \in \mathbf{X}(\Omega)$.*

We emphasize that assumptions $\beta_j \leq \alpha_j \leq 2\beta_j$, $j = 1, 2, 3$, are necessary for lower semi-continuity of $\mathcal{F}$.

The pair (α_j, β_j) provides constrast parameters, possibly different for the various channels, for $j = 1, 2, 3$.

Theorem 4.2. (Existence of strong segmentation for RGB color images)
Let $\Omega \subset \mathbf{R}^2$ be an open set and assume (16) with $\mathbf{g} \in \left(L^{2q}_{loc}(\Omega) \right)^3$. Then there is at least one minimizing triplet $(\mathbf{K}_0, \mathbf{K}_1, \mathbf{u})$ with finite energy for the functional $\mathbf{F}$ defined in (17), with $K_0^j, K_1^j \subset \mathbf{R}^2$ Borel sets, $K_0^j \cup K_1^j$ closed, u_j approximately continuous on $\Omega \setminus K_0^j$ and $u_j \in C^2 \left(\Omega \setminus (K_0^j \cup K_1^j) \right)$. Moreover the sets $K_0^j \cap \Omega$ and $K_1^j \cap \Omega$ are $(\mathcal{H}^1, 1)$ rectifiable.

If $\mathbf{w}$ is a minimizer for $\mathcal{F}$, then a minimizing triplet for $\mathbf{F}$ is given by

$$K_0^j = \overline{S}_{w_j} \setminus \left(S_{\nabla w_j} \setminus S_{w_j} \right), \qquad K_1^j = \overline{S}_{\nabla w_j} \setminus S_{w_j}, \qquad u_j = \widetilde{w}_j \text{ in } \Omega \setminus (K_0^j \cup K_1^j).$$

5. Improvement of the sufficient condition

By inspection of the proof of Theorem 2.6 we can weaken the additional assumption $g \in L^{2q}_{loc}(\Omega)$ in Theorems 2.6 as follows.

Theorem 5.1. *Assume (3), $n = 2$ and there exists a closed set Z such that*

$$\mathcal{H}^1(Z) = 0 \quad and \quad \forall T \subset\subset \Omega \setminus Z \quad \lim_{\rho \to 0+} \sup_{\mathbf{x} \in T} \int_{B_\rho(\mathbf{x})} |g|^{2q} \, d\mathbf{y} = 0 \ . \qquad (19)$$

Then there exists a minimizing triplet for the functional F defined by (1).

Proof. The assumption $g \in L^{2q}_{loc}(\Omega)$ plays a role only in Theorem 5.4 (Decay) of [8] whose proof works in the same way if the assumption is replaced by (19). □

Analogously, the assumption $\mathbf{g} \in \left(L^{2q}_{loc}(\Omega) \right)^3$ in Theorem 4.2 can be weakened as follows.

Theorem 5.2. *Assume (16), $n = 2$ and there exists a closed set Z such that*

$$\mathcal{H}^1(Z) = 0 \text{ and } \quad \forall T \subset\subset \Omega \setminus Z \quad \lim_{\rho \to 0+} \sup_{\mathbf{x} \in T} \int_{B_\rho(\mathbf{x})} |\mathbf{g}|^{2q} \, d\mathbf{y} = 0 \ . \qquad (20)$$

Then there exists a minimizing triplet for the functional **F** *defined by (17).*

References

[1] E. Almansi, *Sull'integrazione dell'equazione differenziale* $\Delta^{2n} = 0$ Ann. Mat. Pura Appl., **III** (1899), 1–51.

[2] L. Ambrosio, N. Fusco & D. Pallara, *Functions of bounded variations and free discontinuity problems,* Oxford Mathematical Monographs, Oxford University Press, Oxford, 2000.

[3] A. Blake & A. Zisserman, *Visual Reconstruction,* The MIT Press, Cambridge, 1987.

[4] M. Carriero, A. Leaci, D. Pallara & E. Pascali, *Euler Conditions for a Minimum Problem with Free Discontinuity Surfaces.* Preprint Dip. di Matematica n. 8, Lecce (1988).

[5] M. Carriero, A. Leaci & F. Tomarelli, *Special Bounded Hessian and elastic-plastic plate,* Rend. Accad. Naz. delle Scienze (dei XL), (109) **XV** (1992), 223–258.

[6] M. Carriero, A. Leaci & F. Tomarelli, *Free gradient discontinuities,* in "Calculus of Variations, Homogeneization and Continuum Mechanics", Buttazzo, Bouchitté, Suquet Eds., World Scientific, Singapore, 1994, 131–147.

[7] M. Carriero, A. Leaci & F. Tomarelli, *A second order model in image segmentation: Blake & Zisserman functional,* in Variational Methods for Discontinuous Structures, R. Serapioni, F. Tomarelli Eds., Birkhäuser, 1996, 57–72.

[8] M. Carriero, A. Leaci & F. Tomarelli, *Strong minimizers of Blake & Zisserman functional,* Ann. Scuola Normale Sup. Pisa, s. IV, **25** (1997), 257–285.

[9] M. Carriero, A. Leaci & F. Tomarelli, *Density estimates and further properties of Blake & Zisserman functional,* in "From Convexity to Nonconvexity", P.D. Panagiotopoulos, R. Gilbert & Pardalos Eds., Nonconvex Optimization and Its Applications 55, Kluwer, 2001, 381–392.

[10] M. Carriero, A. Leaci & F. Tomarelli, *Second order functionals for image segmentation,* Proc. International Workshop "Advanced Mathematical Methods in Electrical and Electronic Measurements", A. Brandolini & F. Tomarelli eds., Esculapio, 2000, 169–179.

[11] M. Carriero, A. Leaci & F. Tomarelli, *Necessary conditions for extremals of Blake & Zisserman functional,* to appear in Comptes Rend. Acad. Sci. Paris, (2002).

[12] M. Carriero, A. Leaci & F. Tomarelli, *Euler equations for Blake & Zisserman functional,* to appear.

[13] E. De Giorgi, *Free discontinuity problems in calculus of variations,* Frontiers in Pure & Applied Mathematics, R. Dautray Ed., North-Holland, Amsterdam, 1991, 55–61.

[14] E. De Giorgi, M. Carriero & A. Leaci, *Existence theorem for a minimum problem with free discontinuity set,* Arch. Rational Mech. Anal., **108** (1989), 195–218.

Michele Carriero
Antonio Leaci
Dipartimento di Matematica "Ennio De Giorgi"
Università di Lecce
Via Arnesano
I-73100 Lecce
Italia
E-mail address: carriero@ingle01.unile.it
E-mail address: leaci@ingle01.unile.it

Franco Tomarelli
Dipartimento di Matematica "Francesco Brioschi"
Politecnico di Milano
Piazza Leonardo da Vinci 32
I-20133 Milano
Italia
E-mail address: fratom@mate.polimi.it

Progress in Nonlinear Differential Equations
and Their Applications, Vol. 51, 81–90

Irrigation

Vicent Caselles and Jean-Michel Morel

Abstract. In many natural or artificial flow systems, a fluid flow network succeeds in connecting every point of a volume to a source. Examples are the blood vessels, the bronchial tree and many irrigation and draining systems. We discuss the irrigation phenomenon from a mathematical point of view and we prove the existence of maximal irrigable sets with minimal cost.

1. Introduction

The function of many natural flow systems is to connect by a fluid flow a finite size volume to a source. This happens, e.g., with the lungs [5]. Typically, the network system is designed according to the following principles: (i) The network supplies an entire volume of the organism and a space filling hierarchical branching pattern is required; (ii) the biological networks have evolved to minimize energy dissipation. Sometimes a third and a fourth principles are added, namely (iii) the size of the final branches of the network is a size-invariant unit, and (iv) the equality of flow supply through the network system [8], [9], [10], [1]. To be able to derive conclusions from this set of principles a basic assumption is usually made, namely that the network has a branched tree structure made of tubes of a certain length, radius and with a given branching number. Then the above principles have been shown to imply that the network has a fractal-like structure with self-similar properties. The network is described by how the branching ratios, and the ratios of radii and lengths of the tubes change through the network. The above principles permit to conclude that the branching ratios are constant, say n, and the radii and length ratios scale as powers of n. This heuristic reasoning ends up with a structure described as a self-similar fractal [8], [9], [10]. Let us mention that in geomorphology, a study of the fractal-like behavior of natural drainage networks was started by R.E. Horton [2], A.N. Strahler [6], and generalized by E. Tokunaga [7].

The above treatment has some weak points, namely, the assumption of the existence of a network structure doing the job, and the assumption that the network is a tree. Both properties should be deduced from first principles, a basic variational principle related to the cost of irrigation should be at the basis of both facts, as requested in [8]. It is our purpose here to discuss the first of them, in particular to show the feasibility of an irrigation system which is an open set containing the source and draining a finite and positive volume, the paths from the source to the irrigated points having bounded length, in other words, with finite length channels. Moreover, we show the possibility of existence of a maximal irrigated volume with a minimal cost for a fairly general class of cost functions.

What remains to be explained is the structure of the maximally irrigated set. An open set irrigating a set of positive measure "must" have a tree-like structure, but we have not made precise what this means, neither described it as a fractal. We do not know if this follows from its optimality. Our plan is as follows. In Section 2 we define irrigable sets and construct some examples of them, based on the construction of a binary fractal tree given in [4]. In Section 3 we prove the existence of maximal irrigable sets. In Section 4 we define cost functions and show the existence of maximal irrigable sets with minimal cost.

2. Irrigable sets

Let $f : [0, \infty) \to [0, \infty)$ be an increasing continuous function such that $f(0) = 0$.

Definition 2.1. *Let U be an open and bounded connected set in $\mathbb{R}^N$, K be a subset of U and $S \in U \setminus K$, $L > 0$. We say that $x \in U$ is L-accessible from S in $U \setminus K$ if there is a curve $\gamma : [0, L(\gamma)] \to \mathbb{R}^N$ parameterized by its arclength such that $\gamma(0) = x$, $\gamma(L(\gamma)) = S$,*

$$(\gamma(s) + B(0, f(s))) \cap K = \emptyset \tag{1}$$

for all $s \in (0, L(\gamma)]$ and $L(\gamma) \le L$. If $x \in U$ and γ is a curve satisfying the above properties, γ will be called a curve of accessibility for x in $U \setminus K$.

For the rest of this section we shall fix U, $S \in U$ and $L > 0$. Given $K \subseteq U$, instead of saying that $x \in U$ is L-accessible from S in $U \setminus K$ we shall simply say that x is accessible in $U \setminus K$ or, simply, accessible.

If γ is an accessibility curve for x in $U \setminus K$, when convenient, we shall extend γ to $[0, L]$ by defining $\gamma(s) = S$ for $s \in (L(\gamma), L]$.

Definition 2.2. *We say that $K \subseteq U$ is irrigable if all points of K are accessible in $U \setminus K$.*

For simplicity we assume in the next lemma that $N = 3$.

Lemma 2.3. *Assume that*

$$\frac{1}{a^3} \sum_{n=1}^{\infty} 2^n f(\frac{a}{2^{n/3}})^3 \to 0 \quad \text{as } a \to 0+, \text{ and} \tag{2}$$

$$\frac{1}{a^2} \sum_{n=1}^{\infty} 2^{2n/3} f(\frac{a}{2^{n/3}})^2 \to 0 \quad \text{as } a \to 0+. \tag{3}$$

Then there is an irrigable subset of U of $\mathbb{R}^3$ with positive measure whose paths have all equal length and profile.

Proof. Without loss of generality we may assume that U contains the origin of coordinates in $\mathbb{R}^3$. The construction of the draining network is based on the construction of a binary fractal tree given in [4]. Let us consider a hyperrectangle in $\mathbb{R}^3$ centered at the origin of coordinates $Q^0 = [-\frac{\alpha}{2}, \frac{\alpha}{2}] \times [-\frac{\beta}{2}, \frac{\beta}{2}] \times [-\frac{\gamma}{2}, \frac{\gamma}{2}]$. We choose $\alpha, \beta, \gamma > 0$ so that $[0, \frac{\alpha}{2}] \times [-\frac{\beta}{2}, \frac{\beta}{2}] \times [-\frac{\gamma}{2}, \frac{\gamma}{2}]$ is similar to Q^0 after the rotation given by $x' = y$, $y' = z$, $z' = x$. This requires that $\alpha : \beta : \gamma$ are in proportions given by $1 : \frac{1}{2^{1/3}} : \frac{1}{2^{2/3}}$. Thus we may take $\alpha = \ell$, $\beta = \frac{\ell}{2^{1/3}}$, $\gamma = \frac{\ell}{2^{2/3}}$, $\ell > 0$. In order to construct the irrigable set we construct the complementary set, i.e., the set of channels irrigating it. We shall proceed iteratively and it will be sufficient to give in detail the first steps of the process. At level 0 we have the hyperrectangle Q^0. At level 1 we have two hyperrectangles Q_1^1, Q_2^1 obtained by subdivision of Q^0 with the plane Oyz (see Figure). At level 2 we have four hyperrectangles Q_i^2, $i = 1, 2, 3, 4$ obtained by subdivision of the hyperrectangles at level 1 with planes parallel to the Oxz plane. At level 3 we subdivide the hyperrectangles at level 2 with planes parallel to the Oxy plane. At level 4 we subdivide the hyperrectangles of level 3 again by planes parallel to the Oyz plane, and we continue iteratively in this way. Thus, at level k we have a family $\mathcal{C}_k$ made of 2^k hyperrectangles whose largest side has length $\alpha_k = \frac{\ell}{2^{k/3}}$.

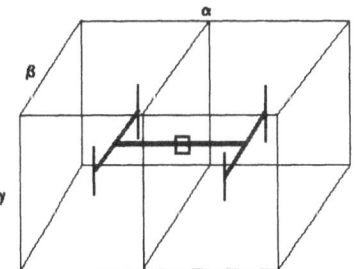

FIGURE 1. A parallepiped similar to its half: a self similar dyadic irrigation tree is easily built in it by connecting the center of the parallepided to the centers of both similar subparallepipeds, and so on.

For each hyperrectangle $Q \in \mathcal{C}_k$ we take the open cube Δ_Q centered at the center of Q and of sidelength δ_k, $k = 0, 1, \ldots$. Given a hyperrectangle $Q \in \mathcal{C}_k$ and its two sons Q_1, Q_2 obtained by subdivision of it, we join the parallel faces of Δ_Q and Δ_{Q_i}, $i = 1, 2$ which are in front of each other by a truncated open pyramid and we do this for all cubes at all levels. These pyramids have volumes

$$\frac{L_k}{3} \frac{\delta_k^3 - \delta_{k-1}^3}{\delta_k - \delta_{k-1}} = \frac{L_k}{3}(\delta_k^2 + \delta_k \delta_{k-1} + \delta_{k-1}^2) \quad \text{where} \quad L_k = \frac{\alpha_{k-1}}{4} - \frac{\delta_{k-1} + \delta_k}{2}.$$

The small cubes Δ_Q and the truncated pyramids constitute a network $\mathcal{N}$ whose volume is

$$V = \sum_{k=0}^{\infty} 2^k \delta_k^3 + \sum_{k=1}^{\infty} 2^k \frac{L_k}{3}(\delta_k^2 + \delta_k \delta_{k-1} + \delta_{k-1}^2). \tag{4}$$

We need that $V < \frac{\ell^3}{2}$ so that its complement $K = Q^0 \setminus \mathcal{N}$ is of positive measure. We shall guarantee that $V < \frac{\ell^3}{2}$ and the set K is irrigable by properly choosing the values of δ_k and ℓ. Observe that any point p in K can be joined to the center of Q^0 by a polygonal curve going through the middle axis of the draining network. The length of the path $\gamma(s)$ from p to the center of Δ_Q, $Q \in \mathcal{C}_n$, is $\frac{1}{4}\sum_{k=n}^{\infty} \alpha_k$. To have irrigability in the sense of Definition 2.2 we must prove that $B(\gamma(s), f(s)) \subseteq N$, which is guaranteed if

$$f(\frac{1}{4}\sum_{k=n}^{\infty} \alpha_k) \le \frac{\delta_{n+1}}{2}.$$

Since

$$\frac{1}{4}\sum_{k=n}^{\infty} \alpha_k = \frac{1}{4}\sum_{k=n}^{\infty} \frac{\ell}{2^{k/3}} = \frac{1}{4}\left(\frac{1}{2^{1/3}-1}\right)\left(\frac{\ell}{2^{(n-1)/3}}\right)$$

the above inequality is implied by the choice

$$\frac{\delta_n}{2} = f(c\frac{\ell}{2^{(n-1)/3}}),$$

where $c = \frac{1}{4}\frac{1}{2^{1/3}-1}$. Then, assuming that $\delta_k \le \delta_{k-1}$, we have

$$\begin{aligned}
V &= \sum_{k=0}^{\infty} 2^k \delta_k^3 + \sum_{k=1}^{\infty} 2^k \frac{L_k}{3}(\delta_k^2 + \delta_k \delta_{k-1} + \delta_{k-1}^2) \\
&\le \sum_{k=0}^{\infty} 2^k \delta_k^3 + \sum_{k=1}^{\infty} 2^k L_k \delta_{k-1}^2 \\
&\le \sum_{k=0}^{\infty} 2^k \delta_k^3 + \frac{1}{4}\sum_{k=1}^{\infty} 2^k \alpha_{k-1}\delta_{k-1}^2 \\
&\le \sum_{k=0}^{\infty} 2^k \delta_k^3 + \frac{1}{2}\sum_{k=0}^{\infty} 2^k \alpha_k \delta_k^2 \\
&\le \sum_{k=0}^{\infty} 2^k \delta_k^3 + \frac{1}{2}\sum_{k=0}^{\infty} 2^{2k/3}\ell\delta_k^2 \\
&\le \sum_{k=0}^{\infty} 2^k 2^3 f(c\frac{\ell}{2^{(k-1)/3}})^3 + \frac{1}{2}\sum_{k=0}^{\infty} 2^{2k/3}\ell f(c\frac{\ell}{2^{(k-1)/3}})^2.
\end{aligned}$$

By our assumptions (2) and (2) on f, by choosing ℓ small enough, the last expression can be made $< \frac{\ell^3}{2}$. Thus we have guaranteed both irrigability of K and its positive measure. $\qquad\square$

Remark 2.4. *Of course the above construction can be adapted to any other dimension $N \geq 2$. Example of functions f satisfying (2) and (3) are $f(s) = s^p$, and $f(s) = \frac{s}{|\log s|^\beta}$, with $p > 1$, $\beta > \frac{1}{3}$. There cannot be irrigable sets of positive measure when f is bounded from below by a linear profile. Indeed, suppose that there is a constant $k > 0$ such that $f(r) \geq kr$ for all $r \in [0, R]$, $R > 0$. Let $L > 0$. Suppose that K be an irrigable set of positive measure and let $O = U \setminus K$. Let $p \in K$ be a point of density 1 in K. Let $\gamma : [0, L(\gamma)] \to \mathbb{R}^N$ be a curve parameterized by its arclength such that $\gamma(0) = p$, $\gamma(L(\gamma)) = S$, and*

$$(\gamma(s) + B(0, f(s))) \cap K = \emptyset$$

for all $s \in (0, L(\gamma)]$, $L(\gamma) \leq L$. Without loss of generality we may assume that $k \leq 1$. Observe that

$$\gamma(\tfrac{r}{2}) + B(0, k\tfrac{r}{2}) \subseteq \gamma(\tfrac{r}{2}) + B(0, f(\tfrac{r}{2})) \subseteq O$$

and

$$\gamma(\tfrac{r}{2}) + B(0, k\tfrac{r}{2}) \subseteq B(p, r)$$

for r small enough. Hence

$$\frac{|B(p, r) \cap O|}{|B(p, r)|} \geq \frac{|B(0, k\tfrac{r}{2})|}{|B(p, r)|} \geq \left(\frac{k}{2}\right)^N$$

for r small enough, and, therefore, $\overline{D}(p, O) \geq (\frac{k}{2})^N$, a contradiction, since p has density 1 in K.

3. Existence of maximally irrigable sets

Given a sequence of subsets K_n of $\mathbb{R}^N$ we define $\limsup K_n = \cap_m \cup_{n \geq m} K_n$, $\liminf K_n = \cup_m \cap_{n \geq m} K_n$.

Proposition 3.1. *Let K_n be a sequence of irrigable sets. There is a subsequence K_{n_j} such that $\limsup K_{n_j}$ is irrigable.*

To prove this result we shall need the following lemma.

Lemma 3.2. *Let K_n be irrigable sets, $x_n \in U$, $x_n \to x$. Let γ_n be an accessibility curve for x_n in $U \setminus K_n$. Then there is a subsequence of K_n which we denote again by K_n and an accessibility curve γ for x in $U \setminus \limsup K_n$. Moreover, there is a parameterization $\tilde{\gamma}$ of γ such that $\gamma_n \to \tilde{\gamma}$ uniformly in $[0, L]$.*

Proof. Since the lengths of the curves γ_n are uniformly bounded (by L), we may assume that γ_n converge uniformly to a curve γ such that $L(\gamma) \leq L$. Indeed, by defining $\gamma_n(s) = S$ when $s \in (L(\gamma_n), L]$ we have that γ_n is uniformly bounded in $W^{1,1}([0, L])$ and we may assume that $\gamma_n \to \tilde{\gamma}$ uniformly in $[0, L]$. Moreover, for each $a \in [0, L]$

$$L(\tilde{\gamma}|_{[0,a]}) \leq \liminf L(\gamma_n|_{[0,a]}) = a$$

Let $\tau(s) = L(\tilde{\gamma}|_{[0,s]})$, then τ maps $[0, L]$ onto $[0, L(\tilde{\gamma})]$. Let γ be the arclength parameterization of $\tilde{\gamma}$. Let $\epsilon > 0$, $0 < \lambda < 1$. Given $\tau_0 \in [0, L(\tilde{\gamma})]$ there is some $s_0 \in [0, L]$, $s_0 \geq \tau_0$, such that $\tau(s_0) = \tau_0$. Then, for n large enough, we have

$$
\begin{aligned}
(\gamma(\tau_0) + B(0, \lambda f(\tau_0))) \setminus B(x, \epsilon) &= (\tilde{\gamma}(s_0) + B(0, \lambda f(\tau_0))) \setminus B(x, \epsilon) \\
&\subseteq (\gamma_n(s_0) + B(0, f(\tau_0))) \setminus B(x, \epsilon) \\
&\subseteq (\gamma_n(s_0) + B(0, f(s_0))) \\
&\subseteq U \setminus K_n.
\end{aligned}
$$

Thus, we have

$$
\gamma(\tau_0) + B(0, f(\tau_0))) \subseteq \cup_m \cap_{n \geq m} U \setminus K_n = U \setminus \limsup K_n
$$

for all $\tau_0 \in [0, L(\gamma)]$, i.e., γ is an accessibility curve for x in $U \setminus \limsup K_n$. $\qquad\square$

Given $A, B \subseteq \mathbb{R}^N$ we define

$$
d(A, B) = \sup_{b \in B} \inf_{a \in A} d(a, b) = \sup_{b \in B} d(A, b).
$$

Proof of Proposition 3.1. Let $x_1 \in \limsup K_n$. Then there is a subsequence K_n^1 of K_n such that $x_1 \in K_n^1$ for all n. Let γ_n be an accessibility curve for x^1 in $U \setminus K_n^1$. By Lemma 3.2, we may assume that $x_1 \in \limsup K_n^1$ and is accessible in $U \setminus \limsup K_n^1$. Let $x_2 \in K^1 := \limsup K_n^1$ be such that

$$
d(x_1, x_2) \geq \frac{d(x_1, (K^1)^c)}{2}.
$$

Again, there is a subsequence K_n^2 of K_n^1 such that $x_2 \in K_n^2$ for all n. Let $K^2 = \limsup K_n^2$. As above, we may assume that x_1, x_2 are accessible in $U \setminus K^2$. Now, we choose $x_3 \in K^2$ such that

$$
d(\{x_1, x_2\}, x_3) \geq \frac{d(\{x_1, x_2\}, (K^2)^c)}{2}.
$$

Iteratively we construct subsequences K_n^i of K_n^{i-1}, sets $K^i = \limsup K_n^i \subseteq K^{i-1}$, and a sequence of points x_i such that $x_1, \ldots, x_i$ are accessible in $U \setminus K^i$, $x_{i+1} \in K^i$, and

$$
d(\{x_1, \ldots, x_i\}, x_{i+1}) \geq \frac{d(\{x_1, \ldots, x_i\}, (K^i)^c)}{2}.
$$

By a diagonal procedure, we find a subsequence K_n' of K_n such that, if $K = \limsup K_n'$, then $\{x_i\}_i \subseteq K$ and each x_i is accessible in $U \setminus K$. By construction, we have

$$
d(\{x_i\}_i, K) = 0,
$$

hence, $\{x_i\}_i$ is dense in K. Again, by Lemma 3.2, we conclude that any point $x \in K$ is accessible in $U \setminus K$.

Lemma 3.3. *If K is irrigable, then $\overline{K}$ is also irrigable. If $O = U \setminus \overline{K}$, then $\overline{K} \subseteq \partial O$.*

Proof. Let $x_n \in K$, $x_n \to x$. Since x_n are accessible in $U \setminus K$, by Lemma 3.2, x is accessible in $U \setminus K$ by a curve γ such that

$$\gamma(s) + B(0, f(s)) \subseteq U \setminus K$$

where s denotes the arclength parameterization of γ, $s \in (0, L(\gamma)]$. Since the set $\{\gamma(s) + B(0, f(s)) : s \in (0, L(\gamma)]\}$ is open, then also

$$\gamma(s) + B(0, f(s)) \subseteq \text{int}(U \setminus K) = U \setminus \overline{K},$$

hence x is accessible in $U \setminus \overline{K}$. Obviously $\overline{K} \subseteq \partial O$. $\square$

From now on, we shall assume that f is such that it exists an irrigable set of positive measure.

Proposition 3.4. *Let* $I := \sup\{|K| : K \text{ is irrigable }\}$. *Then* $0 < I < |U|$ *and there is an irrigable set* K *such that* $|K| = I$.

Proof. By Lemma 2.3, $I > 0$. Let K_n be a sequence of irrigable sets such that $|K_n| \to I$. By Proposition 3.1, we may assume that $K := \limsup K_n$ is irrigable. Then $|K| = I$. Since any point of K is accessible from S, we deduce that $U \setminus K$ contains an open set, hence $I < |U|$. $\square$

4. Maximal irrigable sets at minimal cost

Let H be a functional defined on the set of rectifiable curves, which is lower semicontinuous in the following sense: if $\gamma_n : [0, 1] \to \mathbb{R}^N$ is a sequence of curves with uniformly bounded length such that $\gamma_n \to \gamma$, then $H(\gamma) \le \liminf_n H(\gamma_n)$. Let K be an irrigable set. We define the cost function

$$c_K(x) = \inf_{\gamma \in \mathcal{A}_x} H(\gamma),$$

where $\mathcal{A}_x$ denotes the family of curves of accessibility for x in $U \setminus K$. If $\mathcal{A}_x$ is empty, we set $c_K(x) = 0$.

Lemma 4.1. *Let* K_n *be irrigable sets,* x_n *accessible in* $U \setminus K_n$, $x_n \to x$. *Then there is a subsequence* K_{n_j} *of* K_n *such that* x *is accessible in* $U \setminus K$, *where* $K = \limsup K_{n_j}$, *and* $c_K(x) \le \liminf c_{K_{n_j}}(x_{n_j})$. *As a consequence, if* K *is irrigable,* x_n *are accessibility points in* $U \setminus K$ *and* $x_n \to x$, *then* $c_K(x) \le \liminf c_K(x_n)$.

Proof. Let $\eta > 0$. For each n, let γ_n be an accessibility curve for x_n such that

$$c_{K_n}(x_n) \ge H(\gamma_n) - \eta.$$

By Lemma 3.2, there is a subsequence K_{n_j} of K_n, an accessibility curve γ for x in $U \setminus K$ where $K = \limsup K_{n_j}$ and a parameterization $\tilde{\gamma}$ such that $\gamma_{n_j} \to \tilde{\gamma}$ uniformly in $[0, L]$. Now,

$$c_K(x) \le H(\gamma) \le \liminf_j H(\gamma_{n_j}) \le \liminf_j c_{K_{n_j}}(x_{n_j}) + \eta.$$

Thus, $c_K(x) \le \liminf_j c_{K_{n_j}}(x_{n_j})$. $\square$

To each irrigable set K we associate its irrigation cost defined by

$$E(K) = \int_K c_K(x)\, dx.$$

Let $\mathcal{I}$ be the class of irrigable sets K such that $|K| = I$. We consider the variational problem

$$\text{Min}_{K \in \mathcal{I}}\, E(K). \tag{5}$$

Recall that a dyadic cube is a cube whose vertices are dyadic rationals. For any $k \geq 1$, we shall denote by C_i^k the dyadic cubes whose side has length 2^{-k} and which intersect U. The family of those dyadic cubes will be denoted by $\mathcal{D}^k$.

Theorem 4.2. *The irrigation cost admits a minimum K in $\mathcal{I}$. Moreover, K can be taken as a closed set.*

Proof. The proof follows the lines of an argument given in [3] Let $K_n \in \mathcal{I}$ be such that

$$E(K_n) \to \text{Inf}_{K \in \mathcal{I}}\, E(K).$$

By Proposition 3.1, by extracting a subsequence, we may assume that $K = \limsup K_n$ is irrigable. Notice that $|K| \geq \limsup_n |K_n|$, hence $|K| = I$ since $|K_n| = I$. Notice also that Notice also that $K \cap C = \limsup_n K_n \cap C$, hence $|K \cap C| \geq \limsup_n |K_n \cap C|$ for any dyadic cube $C \in \mathcal{D}^k$. Thus

$$|K| = \lim_n |K_n| \leq \sum_{i=1}^{2^k} \limsup_n |K_n \cap C_i^k| \leq \sum_{i=1}^{2^k} |K \cap C_i^k| = |K|,$$

and, by extracting subsequences, we may assume that $|K_n \cap C| \to |K \cap C|$ as $n \to \infty$ for any dyadic cube $C \in \mathcal{D}^k$. Let

$$M^n(C) = \frac{1}{|K_n \cap C|} \int_{K_n \cap C} c_{K_n}(x)\, dx \qquad C \in \mathcal{D}^k,\ k \geq 1,$$

if $|K_n \cap C| > 0$, $M^n(C) = 0$, otherwise. Suppose that we have extracted a subsequence of K_n such that $M^n(C)$ converges to a limit, called $M(C)$, for any $C \in \mathcal{D}^j$ and any $j \leq k$. Let $C \in \mathcal{D}^{k+1}$ and let $C_1, \ldots, C_{2^N}$ be the 2^N cubes in $\mathcal{D}^k$ such that $C = \cup_{i=1}^{2^N} C_i$. Since

$$M^n(C) = \sum_{i=1}^{2^N} |K_n \cap C_i| M^n(C_i)$$

by extracting a further subsequence of K_n we may assume that $M^n(C_i)$ converge to a value called $M(C_i)$ such that

$$M(C) = \sum_{i=1}^{2^N} |K \cap C_i| M(C_i), \tag{6}$$

and this holds for all cubes $C \in \mathcal{D}^{k+1}$ with their sons C_i, $i = 1, \ldots, 2^N$. By a diagonal extraction procedure, we may assume that $M^n(C)$ converges to $M(C)$ for any dyadic cube C, where $M()$ satisfies (6), i.e., M is a dyadic martingale.

Since M is a bounded martingale in L^1, we may assume that M converges a.e. to a function $F \in L^1(U)$ such that

$$\frac{1}{|C|} \int_C F(x)\, dx \leq M(C),$$

for any dyadic cube C. Now, for any dyadic cube C and n such that $|K_n \cap C| > 0$, we choose a point $x^n(C) \in K_n \cap C$ such that

$$c_{K_n}(x^n(C)) \leq M^n(C)$$

Again, by a diagonal extraction procedure and Lemma 4.1, we may assume that, for any n and any dyadic cube C, $x^n(C)$ converges to a point $x(C)$ which is accessible in $U \setminus \limsup K_n$ and we have that

$$c_K(x(C)) \leq \liminf c_{K_n}(x^n(C)), \tag{7}$$

where $K = \limsup K_n$. Observe that $K \in \mathcal{I}$. Let $x \in K$. We may assume that no coordinate of x is a dyadic rational. Let C^j be a decreasing sequence of dyadic cubes containing x. Note that, by Lemma 4.1, we have

$$c_K(x) \leq \liminf_j c_K(x(C^j)).$$

Let $\epsilon > 0$, and j_0 be such that

$$c_K(x) \leq c_K(x(C^j)) + \epsilon \quad \text{for any } j \geq j_0.$$

Let $j \geq j_0$. Then

$$c_K(x) - \epsilon \leq c_K(x(C^j)) \leq \liminf_n c_{K_n}(x^n(C^j)) \leq \liminf_n M^n(C^j) \leq M(C^j).$$

Thus,

$$c_K(x) \leq F(x) \qquad \text{a.e. in } K,$$

and, therefore,

$$E(K) = \int_K c_K(x)\, dx \leq \int_K F(x)\, dx \leq |K| M(K) - \lim_n E(K_n) - \operatorname{Inf}_{K' \in \mathcal{I}} E(K').$$

The set K attains the infimum of the irrigation energy in $\mathcal{I}$.

Finally, observe that, by Lemma 3.3, $\overline{K} \in \mathcal{I}$. Since $|\overline{K} \setminus K| = 0$, we have that

$$E(\overline{K}) = E(K) = \operatorname{Inf}_{K' \in \mathcal{I}} E(K').$$

$\square$

Acknowledgement. We thank Professor Bernard Sapoval for valuable information, documentation and conversations. The first author acknowledges partial support by the PNPGC project, reference BFM2000-0962-C02-01.

References

[1] A. Bejan, and M.R. Errera, *Deterministic Tree Networks for Fluid Flow: Geometry for Minimal Flow Resistance between a Volume and one Point*, Fractals, **5(4)** (1997), 685–695.

[2] R.E. Horton, *Erosional Development of Streams and their Drainage Basins; Hydrophysical Approach to Quantitative Morphology*, Geol. Soc. Am. Bull. **56** (1945), 275.

[3] S. Masnou and J.M. Morel, *Construction of the solutions to the image disocclusion problem*, In preparation.

[4] W.I. Newman, D.L. Turcotte, and A.M. Gabrielov, *Fractal Trees with Side Branching*, Fractals, **5 (4)** (1997), 603–614.

[5] B. Sapoval, *Universalités et Fractales*, Champs **466**, Flammarion, Paris, France, 1997.

[6] A.N. Strahler, *Quantitative Analysis of Watershed Geomorphology*, Am. Geophys. Un. Trans. **38** (1957), 913.

[7] E. Tokunaga, *Consideration on the Composition of Drainage Networks and their Evolution*, Geographical Rep. Tokyo Metro. Univ. **13** (1978).

[8] G.B. West, *The Origin of Universal Scaling Laws in Biology*, Physica A, **263** (1999), 104–113.

[9] G.B. West, J.H. Brown, and B.J. Enquist, *A General Model for the Structure, Function, and Allometry of Plant Vascular Systems*, Preprint (2001).

[10] G.B. West, J.H. Brown, and B.J. Enquist, *A General Model for the Origin of Allometric Scaling Laws in Biology*, Science, **276(4)** (1997), 122–126.

[11] G.B. West, J.H. Brown, and B.J. Enquist, *The Fractal Dimension of Plants*, Preprint (2001).

Vicent Caselles
Departament de Tecnologia
Universitat Pompeu-Fabra
Passeig de Circumvalació 8
E-08003 Barcelona
Spain
E-mail address: vicent.caselles@tecn.upf.es,

Jean-Michel Morel
CMLA, ENS Cachan
61, Av. du Président Wilson
F-94235 Cachan Cedex
France
E-mail address: morel@cmla.ens-cachan.fr

Progress in Nonlinear Differential Equations
and Their Applications, Vol. 51, 91–102

Symmetrization and Functionals Defined on BV

Andrea Cianchi and Nicola Fusco

Abstract. Regularizing effects on BV functions of the symmetric rearrangement operator are studied and are applied to the discussion of the equality cases in the Pólya-Szegő type inequalities.

1. Introduction

Rearrangements of functions, also called symmetrizations, have proved to be a powerful tool in a number of optimization problems arising in analysis and mathematical phisics. One of the main properties enjoyed by rearrangements is probably the so-called *Pólya-Szegő principle*. The standard version of this principle, which has found relevant applications to Sobolev type inequalities in sharp form ([4], [20]) and to isoperimetric estimates for eigenvalues in the theory of elasticity (see e.g. [21]) can be stated as follows. If u is any (nonnegative) compactly supported function from the Sobolev space $W^{1,p}(\mathbf{R}^n)$, $1 \le p < \infty$, then the symmetric rearrangement $u^\star$ of u also belongs to $W^{1,p}(\mathbf{R}^n)$ and

$$\int_{\mathbf{R}^n} |\nabla u^\star|^p \, dx \le \int_{\mathbf{R}^n} |\nabla u|^p \, dx \ . \tag{1}$$

Inequality (1) is classical and well known. Proofs of (1) can be found e.g. in [8] and [20]; extensions and variants appeared in several papers, including [2], [5], [6], [7], [10], [11], [13], [15], [16], [21].

Clearly, equality holds in (1) provided that $u = u^\star$. On the other hand, the question of whether this assertion can be reversed or not, or more precisely, under which assumptions equality in (1) implies that $u = u^\star$, is a quite subtle problem and has been open for a long time. The first contribution to this question goes back to [17], where the case of analytic functions u is taken into account. A subsequent paper [22] deals with C^n functions. A general sharp result for Sobolev functions is however achieved only in [8], where it is shown that if equality holds in (1) for some $p \in (1, \infty)$ and

$$\mathcal{L}^n \big(\{|\nabla u^\star| = 0\} \cap \{0 < u^\star < \operatorname{esssup} u\} \big) = 0 \ , \tag{2}$$

where $\mathcal{L}^n$ denotes the n-dimensional Lebesgue measure, then $u = u^\star$ (see also the recent paper [14] for an alternate proof). In fact, this conclusion is proved in [8] for more general functionals than $\int_{\mathbf{R}^n} |\nabla u|^p \, dx$. Indeed, it is shown that $u = u^\star$ whenever u satisfies (2) and yields equality in

$$\int_{\mathbf{R}^n} A(|\nabla u^\star|) \, dx \leq \int_{\mathbf{R}^n} A(|\nabla u|) \, dx \,, \tag{3}$$

where A is a Young function, namely a convex function from $[0, +\infty)$ into $[0, +\infty)$ vanishing at 0, satisfying the additional assumption that $A^{1/q}$ is still convex for some $q > 1$. Even in this extended version, however, important functionals having linear growth (like the area functional corresponding to $A(s) = \sqrt{1 + s^2} - 1$) or nearly linear growth are ruled out.

The original motivation of the work reported in the present note was to understand whether the class of functionals having linear growth could be included in the discussion of the equality case of (3) and, especially, whether the regularity assumptions on u could be further relaxed to allow even test functions of bounded variations (briefly BV functions), the natural class of functions where functionals of the Calculus of Variations with linear growth can be suitably defined. It turns out that this is possible, but requires a preliminary study of regularizing effects of the symmetric rearrangement operator in connection with the pointwise behaviour of BV functions. Actually, we found the discovery of these regularizing properties the most intriguing part of our research; a description of this part is presented in Section 2. Section 3 is devoted to Pólya-Szegő type inequalities for functionals defined on BV and to the characterization of extremals in these inequalities. For complete proofs of the results, we refer to [12].

2. Regularization under symmetric rearrangement

Let u be any nonnegative, measurable and compactly supported function in $\mathbf{R}^n$. The symmetric rearrangement $u^\star$ of u is a radially decreasing function in $\mathbf{R}^n$ whose level sets have the same measure as the level sets of u. In formulas

$$u^\star(x) = \sup\{t \geq 0 : \mathcal{L}^n(\{x \in \mathbf{R}^n : u(x) > t\}) > \omega_n |x|^n\} \qquad \text{for } x \in \mathbf{R}^n \,,$$

where ω_n denotes the measure of the unit ball in $\mathbf{R}^n$. As expected from this definition, the theory of rearrangements heavily relies on properties of level sets of functions and of their restrictions to their level surfaces. In particular, in dealing with BV functions we are led to work with (essential) boundaries of their level sets rather than with their level surfaces. Recall that a function $u \in L^1(\mathbf{R}^n)$ belongs to the space $BV(\mathbf{R}^n)$ of functions of bounded variation in $\mathbf{R}^n$ if its distributional gradient Du is an $\mathbf{R}^n$-valued Radon measure in $\mathbf{R}^n$ and the total variation $|Du|$ of Du is finite in $\mathbf{R}^n$. Moreover, given any $t > 0$, the essential boundary $\partial^*\{u > t\}$ of the level set $\{u > t\}$ is defined as the complement in $\mathbf{R}^n$ of the set of points where $\{u > t\}$ has density either 1 or 0.

Aim of this section is to clarify to what extent the degree of regularity of $u^\star$ restricted to the boundary of one of its level sets $\{u^\star > t\}$ reflects the regularity of u restricted to the essential boundary $\partial^*\{u > t\}$ of the corresponding level set $\{u > t\}$. The flavor of the result is that the symmetric rearrangement strengthens the pointwise properties of u over $\partial^*\{u > t\}$. Indeed, our conclusions can be summarized in the following principles.

- For a.e. $t > 0$ such that the essential boundary $\partial^*\{u > t\}$ of the level set $\{u > t\}$ contains at least one point of approximate continuity for u, the essential boundary $\partial^*\{u^* > t\}$ of the corresponding level set $\{u^* > t\}$ agrees with the level surface $\{u^* = t\}$ and consists of points of approximate continuity for u^*.

- For a.e. $t > 0$ such that $\partial^*\{u > t\}$ contains a subset of positive $(n-1)$-dimensional Hausdorff measure of points of approximate differentiability for u with non-vanishing approximate gradient, $\partial^*\{u^* > t\}$ agrees with $\{u^* = t\}$ and consists of points of approximate differentiability for u^* with non-vanishing approximate gradient.

A natural question which can be raised is whether these assertions can be reversed or not. Interestingly, the answer is affirmative when $n = 1$, whereas is negative, in general, in dimension $n \geq 2$. Actually, in the latter case, there exist BV functions u whose rearrangement u^* is continuous on every level surface $\{u^* = t\}$, without there being any point of approximate continuity for u on $\partial^*\{u > t\}$, and there exist BV functions u whose rearrangement u^* is smooth with nonzero gradient on a.e. level surface $\{u^* = t\}$, without there being a set of positive $(n-1)$-dimensional Hausdorff measure of points of approximate differentiability for u with non-vanishing approximate gradient on $\partial^*\{u > t\}$.

A formal statement of these facts requires a few more notations. Given a real-valued measurable function u in $\mathbf{R}^n$, we denote by C_u the set of points of approximate continuity of u and by S_u its complement in $\mathbf{R}^n$. Moreover, we call $\mathcal{D}_u$ the set of those points in C_u where u admits an approximate gradient ∇u, and by $\mathcal{D}_u^+$ and $\mathcal{D}_u^0$ the subsets of $\mathcal{D}_u$ where $\nabla u \neq 0$ and $\nabla u = 0$, respectively. As usual, $\mathcal{H}^k$ stands for the k-dimensional Hausdorff measure. We begin with the result about the points of approximate continuity.

Theorem 2.1. *Let u be a nonnegative, measurable and compactly supported function in $\mathbf{R}^n$.*
Part I.

$$\emptyset \neq \{u^* = t\} \subset C_{u^*} \quad \textit{for a.e. } t > 0 \quad \textit{such that } \partial^*\{u > t\} \cap C_u \neq \emptyset. \quad (4)$$

Part II. Assertion (4) cannot be reversed. Indeed, there exist functions u as above such that

$$\emptyset \neq \{u^* = t\} \subset C_{u^*} \quad \textit{for every } t > 0 \quad \textit{but } \partial^*\{u > t\} \cap C_u = \emptyset \quad \textit{for a.e. } t > 0. \quad (5)$$

Assume, in addition, that $u \in BV(\mathbf{R}^n)$. If $n = 1$, the converse of (4) is true; namely,

$$\partial^*\{u > t\} \cap C_u \neq \emptyset \quad \textit{for a.e. } t > 0 \quad \textit{such that } \emptyset \neq \{u^* = t\} \subset C_{u^*}. \quad (6)$$

If $n \geq 2$, assertion (6) can be false. Indeed, functions $u \in BV(\mathbf{R}^n)$ exist satisfying (5).

The points of approximate differentiability are dealt with in the next theorem.

Theorem 2.2. *Let u be a nonnegative compactly supported function from $BV(\mathbf{R}^n)$.*
Part I.

$$\{u^* = t\} \subset \mathcal{D}_{u^*}^+ \quad \text{for a.e. } t > 0 \quad \text{such that } \mathcal{H}^{n-1}\left(\partial^*\{u > t\} \cap \mathcal{D}_u^+\right) > 0.$$

Part II. Assume, in addition, that either $n = 1$, or $n \geq 2$ and

$$\mathcal{L}^n(\mathcal{D}_u^0 \cap \{0 < u < \operatorname{esssup} u\}) = 0. \tag{7}$$

Then

$$\mathcal{H}^{n-1}\left(\partial^*\{u > t\} \cap \mathcal{D}_u^+\right) > 0 \quad \text{for a.e. } t > 0 \quad \text{such that } \{u^* = t\} \subset \mathcal{D}_{u^*}^+. \tag{8}$$

If $n \geq 2$ and assumption (7) is dropped, then assertion (8) can be false. Indeed,
there exist functions $u \in BV(\mathbf{R}^n)$ such that

$$\{u^* = t\} \subset \mathcal{D}_{u^*}^+ \quad \text{for every, but one, } t > 0, \quad \text{but } \mathcal{H}^{n-1}(\partial^*\{u > t\} \cap \mathcal{D}_u^+) = 0 \tag{9}$$

for a.e. $t > 0$.

We do not reproduce full proofs of Theorems 2.1–2.2 here, but we limit ourselves to sketching counterexamples proving the negative assertions contained in
Parts II.

Example 2.3. Here we construct a measurable function $u : \mathbf{R} \to [0, 1]$ satisfying
(5). Consider the sequence of points $a_h \in [0, 1]$ defined as $a_h = 0$ if $h = 0$, and
$a_h = \sum_{j=0}^{h-1} 2^j / 3^{j+1}$ if $h \in \mathbf{N}, h \geq 1$. Subdivide each interval (a_{h-1}, a_h) into the
2^{h-1} intervals whose interiors I_h^j are given by

$$I_h^j = \left(a_{h-1} + \frac{j-1}{3^h}, a_{h-1} + \frac{j}{3^h}\right), \qquad j \in \{1, \ldots, 2^{h-1}\}.$$

Then, we define u by

$$u(x) = \begin{cases} \dfrac{2j-1}{2^h} & \text{if } x \in I_h^j \text{ for some } h \geq 1 \text{ and } j \in \{1, \ldots, 2^{h-1}\} \\ 0 & \text{otherwise}. \end{cases}$$

The function u is measurable, but does not belong to $BV(\mathbf{R})$, since $|Du|((a_{h-1}, a_h))$
$= (2^{h-1} - 1)/2^{h-1}$ for every $h \geq 2$. Moreover, $\partial^*\{u > t\} \cap C_u = \emptyset$ if $t \neq (2j-1)/2^h$
for every $h \in \mathbf{N}$ and $j \in \{1, \ldots, 2^{h-1}\}$. On the other hand, μ_u is not constant in
any interval. Thus, u^* is continuous in $(-1/2, 1/2)$. Hence, $\{u^* = t\} \subset C_{u^*}$ for
every $t > 0$. Incidentally, let us notice that u^* agrees in $(0, 1)$ with the classical
(non-increasing) Cantor-Vitali function, which is everywhere continuous.

Example 2.4. A function $u \in BV(\mathbf{R}^n)$, $n \geq 2$, is produced satisfying (5). For
simplicity of notations, we assume that $n = 2$. It will be then clear that every-
thing carries over to higher dimensions. The construction is similar to that of
Example 2.3.

Consider any family of open squares Q_h^j, where $h \in \mathbf{N}$, $h \geq 1$ and $j \in \{1, \ldots, 2^{h-1}\}$, whose closures are pairwise disjoint, and such that each Q_h^j is contained in the square $(0,1)^2$ and has side-length not exceeding $1/4^h$. We define u as

$$u(x) = \begin{cases} \dfrac{2j-1}{2^h} & \text{if } x \in Q_h^j \text{ for some } h \geq 1 \text{ and } j \in \{1, \ldots, 2^{h-1}\} \\ 0 & \text{otherwise .} \end{cases}$$

As in Example 2.3, the function u is measurable and $\partial^*\{u > t\} \cap C_u = \emptyset$ whenever $t \neq (2j-1)/2^h$ for every $h \geq 1$ and $j \in \{1, \ldots, 2^{h-1}\}$. Furthermore, u^* is continuous, whence $\{u^* = t\} \subset C_{u^*}$ for every $l > 0$. However, unlike the one-dimensional function of Example 2.3, the function u is of bounded variation, since

$$|Du|(\mathbf{R}^2) = \sum_{h=1}^{\infty} \sum_{j=1}^{2^{h-1}} |Du|(\overline{Q}_h^j) < 4 \sum_{h=1}^{\infty} \frac{2^{h-1}}{4^h} < +\infty .$$

Example 2.5. We exhibit a function $u \in BV(\mathbf{R}^n)$, $n \geq 2$, whose support equals the unit cube $[0,1]^n$ and whose range is in $[0,1)$, satisfying (9).

Our example is a variant of that constructed in [23]; let us mention that an adaptation of that example was also considered in [1] for other purposes. As in Example 2.4, for the sake of simplicity we shall carry out computations in dimension $n = 2$.

Denote by Q the square $(0,1)^2$. Divide Q into four congruent squares having sides parallel to those of Q and denote their hinteriors by $Q_0^1, Q_1^1, Q_2^1, Q_3^1$, starting with the square on the top left corner and proceeding clockwise. Define the function $v_1 : \mathbf{R}^n \to [0,1)$ by

$$v_1(x) := \begin{cases} \dfrac{i_1}{4} & \text{if } x \in Q_{i_1}^1 \text{ for some } i_1\{0,1,2,3\} \\ 0 & \text{otherwise .} \end{cases}$$

Now, repeat this construction and subdivide each square $Q_{i_1}^1$ into four squares whose interiors are labeled by $Q_{i_1,0}^2, Q_{i_1,1}^2, Q_{i_1,2}^2, Q_{i_1,3}^2$, starting again with the square on the top left corner and proceeding clockwise. Then define $v_2 : \mathbf{R}^n \to [0,1/4)$ by

$$v_2(x) := \begin{cases} \dfrac{i_2}{4^2} & \text{if } x \in Q_{i_1,i_2}^2 \text{ for some } i_1, i_2 \in \{0,1,2,3\} \\ 0 & \text{otherwise .} \end{cases}$$

On iterating this procedure, at the m-th iteration each of the squares $Q_{i_1,\ldots,i_{m-1}}^{m-1}$ constructed at the $(m-1)$-th iteration is divided into four squares $Q_{i_1,\ldots,i_{m-1},i_m}^m$ and $v_m : \mathbf{R}^n \to [0, 1/4^{m-1})$ is defined by

$$v_m(x) := \begin{cases} \dfrac{i_m}{4^m} & \text{if } x \in Q_{i_1,\ldots,i_{m-1},i_m}^m \text{ for some } i_1, \ldots, i_{m-1}, i_m \in \{0,1,2,3\} \\ 0 & \text{otherwise .} \end{cases}$$

The sequence of functions

$$\{u_m\} \quad \text{defined by} \quad u_m = \sum_{l=1}^{m} v_l \quad \text{for} \quad m \in \mathbf{N}$$

is non-decreasing. It is not difficult to verify that $u \in BV(\mathbf{R}^2)$ and that $\nabla u = 0$ a.e. One can show that $u^\star = \max\{1 - \omega_n |x|^n, 0\}$, whence $\emptyset \neq \{u^\star = t\} \subset \mathcal{D}_{u^\star}^+$ for every $t \in (0,1)$, whereas

$$S_u = \partial Q \cup \left(\cup_m \partial Q_{i_1,\ldots,i_m}^m\right) \setminus \{(0,1)\}$$

and

$$\mathcal{H}^1(\partial^\star \{u > t\} \cap \mathcal{D}_u^+) = 0 \qquad \text{for a.e. } t > 0 .$$

The results of Theorems 2.1–2.2, of a local nature, have consequences at a global level. Let us begin their description by taking into account the total variation of the derivatives of a function. Given $u \in BV(\mathbf{R}^n)$, denote by $D^a u$ the absolutely continuous part of Du and by $D^s u$ its singular part, which, in turn, can be split into the Cantor part $D^c u$ and the jump part $D^j u$. The sum of $D^a u$ and $D^c u$ will be referred to as the diffuse part of Du. The total variation of these measures will be denoted by vertical bars.

It is an easy consequence of the coarea formula and of the isoperimetric inequality that if u is any nonnegative compactly supported function from $BV(\mathbf{R}^n)$, then $u^\star$ is also in $BV(\mathbf{R}^n)$ and the total variation of $Du^\star$ does not exceed that of Du. In Theorem 2.6 below we pursue a closer investigation on this point and analyze the behaviour under rearrangement of the total variations of $D^s u$ and $D^j u$, separately. The result leads us to maintain that the less functions (which are differentiable is some sense) are smooth, the more they benefit in regularity from symmetrization. Indeed, whereas classical differentiability is not inherited by rearranged functions, and weak differentiability is just preserved after rearrangement, Theorem 2.6 tells us that the operation of symmetric rearrangement tends to regularize functions u which are differentiable in the sense of measures by reducing the singular part $D^s u$ of their gradient, and by reducing, in $D^s u$, the most concentrated part $D^j u$.

Theorem 2.6. *Let u be a nonnegative compactly supported function from $BV(\mathbf{R}^n)$. Then $u^\star \in BV(\mathbf{R}^n)$. Moreover,*

$$|Du^\star|(\mathbf{R}^n) \leq |Du|(\mathbf{R}^n) , \tag{10}$$

$$|D^s u^\star|(\mathbf{R}^n) \leq |D^s u|(\mathbf{R}^n) , \tag{11}$$

$$|D^j u^\star|(\mathbf{R}^n) \leq |D^j u|(\mathbf{R}^n) . \tag{12}$$

Notice also that, as shown by simple examples, the total variations of $D^a u$ and $D^c u$ may be enhanced by symmetrization, so that no analogs of (10)–(12) for $|D^a u|(\mathbf{R}^n)$ or $|D^c u|(\mathbf{R}^n)$ can hold true. Consider, for instance, the function $u : \mathbf{R} \to [0,\infty)$ which is defined by $u(x) = 1-x$ if $x \in [0,1]$ and vanishes otherwise. Then $u^\star(x) = 1 - 2|x|$ if $x \in [-\frac{1}{2}, \frac{1}{2}]$ and vanishes otherwise. Thus, $|D^a u|(\mathbf{R}) = 1$ and $|D^s u|(\mathbf{R}) = |D^j u|(\mathbf{R}) = 1$, whereas $|D^a u^\star|(\mathbf{R}) = 2$ and $|D^s u^\star|(\mathbf{R}) = 0$.

Similarly, if $u : \mathbf{R} \to [0, \infty)$ agrees with a purely (non-increasing) Cantor type function in $[0, 1]$ and vanishes elsewhere, then $u^\star(x) = u(2|x|)$ if $x \in [-\frac{1}{2}, \frac{1}{2}]$ and vanishes otherwise. Thus, $|D^c u|(\mathbf{R}) = 1$ and $|D^j u|(\mathbf{R}) = 1$, but $|D^c u^\star|(\mathbf{R}) = 2$ and $|D^j u^\star|(\mathbf{R}) = 0$. In accordance with Theorem 2.6, $D^j u$, the most irregular part of Du, is reduced (and vanishes, in fact) after symmetrization, in favor of the more regular parts $D^a u$ and $D^c u$, in the former and in the latter example, respectively. However, observe that $D^a u$ and $D^c u$, respectively, are already not identically equal to zero on the boundary of every level set of u before symmetrizing. This situation is typical of one-dimensional functions, where, by Theorems 2.1–2.2, Parts II, no absolutely continuous part or diffuse part of $Du^\star$ can be created on $\{u^\star = t\}$ if not already present in Du restricted to $\partial^\star \{u > t\}$. Instead, in higher dimension, functions u such that Du has only a jump part may be mapped, by symmetric rearrangement, into functions whose Du is purely diffuse or even purely absolutely continuous – see Examples 2.4–2.5.

3. Pólya-Szegő type inequalities

A functional on $BV(\mathbf{R}^n)$ can be associated with any given convex function from $\mathbf{R}^n$ into $[0, +\infty)$ growing linearly at infinity. The purpose of this section is to describe the effects of the symmetric rearrangement on those functionals from this class which depend only on $|Du|$.

Let A be any Young function, namely a convex function from $[0, +\infty)$ into $[0, +\infty)$ vanishing at 0, and assume, in addition, that a positive constant k exists such that

$$A(s) \leq k(1 + s) \qquad \text{for } s \geq 0 . \tag{13}$$

Set

$$\lim_{s \to +\infty} \frac{A(s)}{s} = A_\infty . \tag{14}$$

Observe that, by our assumptions on A, the limit in (14) exists and is finite. Then we define the functional

$$J_A : BV(\mathbf{R}^n) \to [0, +\infty)$$

as

$$J_A(u) = \int_{\mathbf{R}^n} A(|\nabla u|) \, dx + A_\infty |D^s u|(\mathbf{R}^n) . \tag{15}$$

The integral on the right-hand side of (15) is well defined, since when $u \in BV(\mathbf{R}^n)$, its approximate gradient ∇u is defined a.e. in $\mathbf{R}^n$ and agrees a.e. with the density of the absolutely continuous part of Du with respect to the Lebesgue measure. Let us emphasize that definition (15) is standard and very natural in the present framework, since the right-hand side of (15) agrees with the relaxed functional in $BV(\mathbf{R}^n)$, with respect to convergence in L^1, of the functional $\int_{\mathbf{R}^n} A(|\nabla u|) \, dx$ defined on C^1 (or $W^{1,1}$) functions u (see [9]). As mentioned in Section 1, a typical example of functionals with linear growth is provided by the so-called area functional corresponding to the choice $A(s) = \sqrt{1 + s^2} - 1$. Indeed, with such a choice, $J_A(u)$ yields (up to the measure of the support of u) the perimeter of the essential boundary of the subgraph of the function u. The following result is a Pólya-Szegő principle for the functional J_A and provides a version of inequality (3) in the framework of BV functions.

Theorem 3.1. *Let A be any Young function satisfying (13). Then*

$$J_A(u^\star) \le J_A(u) \tag{16}$$

for any nonnegative compactly supported function $u \in BV(\mathbf{R}^n)$.

Inequality (16) could be derived, via relaxation, from inequality (3). However, this kind of a proof is of no use in the discussion of the equality sign. Alternatively, a direct proof can be supplied, based on inequalities over the essential boundaries of the level sets of u and $u^\star$. This approach, which relies on a general version of the coarea formula and on the regularity principles of Theorems 2.1–2.2, is an appropriate starting point to characterize extremals in (16). The result is summarized in the next theorem.

Theorem 3.2. *Let A be a Young function satisfying (13). Assume that*

$$J_A(u^\star) = J_A(u) \tag{17}$$

for some nonnegative compactly supported function $u \in BV(\mathbf{R}^n)$.
(i) If A is strictly increasing, then u agrees almost everywhere with a function whose level sets are open balls.
(ii) If, in addition, A is strictly convex, then:
for a.e. $t \in (0, \infty) \setminus u^\star(\mathbf{R}^n)$,

$$\partial^*\{u > t\} \cap C_u = \emptyset; \tag{18}$$

for a.e. $t \in u^\star(\mathbf{R}^n) \setminus u^\star(\mathcal{D}_{u^*}^+)$,

$$\mathcal{H}^{n-1}(\partial^*\{u > t\} \cap \mathcal{D}_u^+) = 0; \tag{19}$$

for a.e. $t \in u^\star(\mathcal{D}_{u^*}^+)$,

$$\mathcal{H}^{n-1}(\partial^*\{u > t\} \setminus \mathcal{D}_u^+) = 0 \tag{20}$$

and

$$|\nabla u(x)| = |\nabla u^\star|_{|\{u^*=t\}} \qquad \text{for } \mathcal{H}^{n-1}\text{-a.e. } x \in \partial^*\{u > t\} . \tag{21}$$

(iii) If, furthermore,

$$\mathcal{L}^n(\mathcal{D}_{u^*}^0 \cap \{0 < u^\star < \operatorname{esssup} u\}) = 0 , \tag{22}$$

then u agrees a.e. with a translate of $u^\star$.

Let us make a fex comments on the hypotheses of Theorem 3.2. The assumption that A is strictly increasing is equivalent to requiring that A is strictly positive in $(0, +\infty)$. This assumption is indispensable, since if $s_0 > 0$ exists such that A vanishes in $[0, s_0]$, then for every Lipschitz continuous function u such that $|\nabla u| \le s_0$ a.e., $u^\star$ is still Lipschitz continuous, $|\nabla u^\star| \le s_0$ a.e. and $J_A(u) = J_A(u^\star) = 0$.

Equations (18) and (19) hold irrespective of whether (17) is satisfied or not, as a consequence of Theorems 2.1 and 2.2, respectively. We have included them in the statement of Theorem 3.2 in parallel with (20)–(21) for completeness. The assumption that A is strictly convex cannot be removed for the remaining assertions in (ii)–(iii) to be true. Actually, if $A(s) = as$ for some $a > 0$, then $J_A(u) = a|Du|(\mathbf{R}^n)$. Hence, $J_A(u^\star) = J_A(u)$ for every function $u \in BV(\mathbf{R}^n)$ whose level sets are (not necessarily concentric) balls. In particular, even if u is smooth, its gradient need not to be constant on the level surfaces of u. Similar counterexamples can be easily exibited even if A is affine only in some subinterval of $[0, +\infty)$.

Condition (22) is implied by the corresponding condition

$$\mathcal{L}^n(\mathcal{D}_u^0 \cap \{0 < u < \operatorname{esssup} u\}) = 0 \qquad (23)$$

on u. The reverse implication is not true in general, as shown, for instance, by Example 2.5. However, (22) and (23) are equivalent for any function $u \in BV(\mathbf{R}^n)$ satisfying (17) for some strictly convex A.

Condition (22) implies that

$$\mathcal{L}^n(\{u = t\}) = 0 \qquad\qquad \text{for every } t \in (0, \operatorname{esssup} u), \qquad (24)$$

and is, in fact, stronger than (24). Actually, (24) is equivalent to the continuity of the distribution function of u, whereas (22) is equivalent to the absolute continuity of such distribution function. If (22) does not hold, then assertion (iii) can be false. This is easily seen if not even (24) is fulfilled. Indeed, consider any spherically symmetric function v having the property that $\mathcal{L}^n(\{v = 1\}) > 0$ and another function u whose graph agrees with the graph of v where $0 \le u \le 1$, and with a (slight) translated of the graph of v where $u > 1$. Obviously, $J_A(u) = J_A(v) = J_A(u^\star)$, but $u \ne v = u^\star$ on a set of positive measure. Finer counterexamples where (24) holds, but (22) does not, are constructed in [8].

We conclude with an outline of the proof of Theorem 3.2. Assertions (i) and (ii) follow from an inspection of the proof of inequality (16) under the additional information that it holds as an equation. The main results to be used here are the characterization of balls as the only sets yielding equality in the classical isoperimetric inequality in $\mathbf{R}^n$, and the characterization of constant functions as the only integrands yielding equality in Jensen's inequality for a strictly convex function.

As far as assertion (iii) is concerned, the proof in the one-dimensional case and the proof in higher dimension are substantially different. The former is considerably easier, but, interestingly enough, it is not just a simplification of the latter. Indeed, the higher-dimensional proof cannot be adapted to fit one-dimensional situations, the major technical obstacle being that any nonempty subset of $\mathbf{R}$ has positive $\mathcal{H}^0$ measure.

Let us focus on the case when $n \ge 2$. Our approach in this case combines ideas from [3], [8] and [14] with ad hoc arguments developed to overcome specific difficulties due to the lack of regularity of the relevant functions. The outline of the proof is as follows.

As a preliminary step we observe that assertion (i) implies that we may always assume, possibly modifying u on a set of zero measure, that for all $t \in [0, \text{esssup}\, u)$ the level set $U_t = \{u > t\}$ is an open ball and that the set $\Lambda = \{u = \text{esssup}\, u\}$ is a closed ball (possibly a single point). Then we start as in [14] and factorize u as

$$u(x) = u^*(\mu_u(u(x))) \qquad \text{for a.e. } x \in \mathbf{R}^n \qquad (25)$$

Equation (25) is straightforward and holds for every $x \in \mathbf{R}^n$ when u is weakly differentiable, since u^* is (absolutely) continuous in that case and hence $u^*(\mu_u(t)) = t$ for every $t \in (0, \text{esssup}\, u)$. In the present framework, only the inequality $u^*(\mu_u(t)) \leq t$ is known, so that (25) requires a proof. The striking fact is however that $\mu_u \circ u$ belongs to the Sobolev space $W^{1,1}(U_0 \setminus \Lambda)$ and that a chain rule holds for $\nabla(\mu_u \circ u)$. This is a delicate point, since, even if μ_u is absolutely continuous, u is just a BV function. In order to get over this obstacle, we call into play a map $T : \mathbf{R}^n \to \mathbf{R}^n$ introduced in [8] which relates u^* with u by the equation $u^*(x) = u(T(x))$ for a.e. $x \in \mathbf{R}^n$ and is defined as follows. Let $\gamma : [0, \text{esssup}\, u] \to \mathbf{R}^n$ be the function given by

$$\gamma(t) = \begin{cases} \text{the center of } U_t & \text{if } 0 \leq t < \text{esssup}\, u \\ \text{the center of } \Lambda & \text{if } t = \text{esssup}\, u, \end{cases}$$

and let $\tau : [0, +\infty) \to \mathbf{R}^n$ be defined as

$$\tau(r) = \begin{cases} \gamma(u^*(\omega_n r^n)) & \text{if } 0 \leq r \leq r_0 \\ 0 & \text{if } r_0 \leq r, \end{cases}$$

where r_0 is the radius of U_0. Then T is given by

$$T(x) = x + \tau(|x|).$$

It turns out that T is invertible outside a set of zero Lebesgue measure and that $\mu_u \circ u$ agrees with $\omega_n |T^{-1}|^n$ a.e. Specific properties of the map T allow us to make use of results from geometric measure theory and infer that $|T^{-1}|$, and hence $\mu_u \circ u$, belongs to $W^{1,1}(U_0 \setminus \Lambda)$. The information on Du contained in (ii) then enables one to express $\nabla(\mu_u \circ u)$ in terms of $\mu \circ u$ itself and of ∇u. The resulting formula tells us that the function $\left(\dfrac{\mu_u \circ u}{\omega_n}\right)^{1/n}$ (has balls as level sets and) agrees a.e. with a Lipschitz function whose gradient has length equal to 1 a.e. An argument from [3] based on the study of the lines of steepest descent then entails that the level sets of $\left(\dfrac{\mu_u \circ u}{\omega_n}\right)^{1/n}$ are concentric, whence by (25), we may conclude that also those of u are.

References

[1] F.J. Almgren and E.H. Lieb, *Symmetric rearrangement is sometimes continuous*, J. Amer. Math. Soc. **2** (1989), 683–773.

[2] A. Alvino, V. Ferone, P.L. Lions and G. Trombetti, *Convex symmetrization and rearrangements*, Ann. Inst. H. Poincaré, Anal. Non Lineaire **14** (1997), 275–293.

[3] G. Aronsson and G. Talenti, *Estimating the integral of a function in terms of a distribution function of its gradient*, Boll. Un. Mat. Ital. (5), **18-B** (1981), 885–894.

[4] T. Aubin, *Problèmes isopérimétriques et éspaces de Sobolev*, J. Diff. Geom. **11** (1976), 573–598.

[5] A. Baernstein II, *A unified approach to symmetrization*, in Partial differential equations of ellyptic type, A. Alvino, E. Fabes and G. Talenti eds., Symposia Math. 35, Cambridge Univ. Press, 1994.

[6] F. Betta, F. Brock, A. Mercaldo and M. Posteraro, *A weighted isoperimetric inequality and applications to symmetrization*, J. Ineq. Appl. **4** (1999), 215–240.

[7] F. Brock, *New Dirichlet-type inequalites for Steiner-symmetrization*, Calc. Var. **8** (1999), 15–25.

[8] J. Brothers and W. Ziemer, *Minimal rearrangements of Sobolev functions*, J. Reine. Angew. Math. **384** (1988), 153–179.

[9] G. Buttazzo, *Semicontinuity, relaxation and integral representation in the calculus of variations*. Pitman Res. Notes Math. Ser., **207**, Longman, 1989.

[10] A. Cianchi, *Second order derivatives and rearrangements*, Duke Math. J. **105** (2000), 355–385.

[11] A. Cianchi, *Rearrangements of functions in Besov spaces*, Math. Nachr. **230** (2001), 19–35.

[12] A. Cianchi and N. Fusco, *Functions of bounded variation and rearrangements*, preprint (2001).

[13] A. Ehrhard, *Inégalités isopérimétriques et intégrales de Dirichlet gaussiennes*, Ann. Sci. Ecole Norm. Sup. **17** (1984), 317–332.

[14] A. Ferone and R. Volpicelli, *Minimal rearrangements of Sobolev functions: a new proof*, Ann. Inst. H. Poincaré, Anal. Non Linéaire, to appear.

[15] A.M. Garsia and E. Rodemich, *Monotonicity of certain functionals under rearrangements*, Ann. Inst. Fourier **24** (1974), 67–116.

[16] B. Kawohl, *Rearrangements and convexity of level sets in PDE*, Lecture Notes in Math. **1150**, Springer-Verlag, Berlin, 1985.

[17] B. Kawohl, *On the isoperimetric nature of a rearrangement inequality and its consequences for some variational problems*, Arch. Rat. Mech. Anal. **94** (1986), 227–243.

[18] V.M. Maz'ya, *Sobolev spaces*, Springer-Verlag, Berlin, 1985.

[19] R. O'Neil, *Convolution operators and L(p,q) spaces*, Duke Math. J. **30** (1963), 129–142.

[20] G. Talenti, *Best constant in Sobolev inequality*, Ann. Mat. Pura Appl. **110** (1976), 353–372.

[21] G. Talenti, *A weighted version of a rearrangement inequality*, Ann. Univ. Ferrara **43** (1997), 121–133.

[22] A. Uribe, *Minima of the Dirichlet norm and Toeplitz operators*, preprint, 1985.

[23] H. Whitney, *A function not constant on a connected component of critical points* , Duke Math. J. **1** (1935), 514–517.

[24] W.P. Ziemer, *Weakly differentiable functions*, Springer-Verlag, New York, 1989.

Andrea Cianchi
Dipartimento di Matematica e Applicazioni per l'Architettura
Università di Firenze
Piazza Ghiberti 27
I-50122 Firenze, Italy
E-mail address: cianchi@mail.unifi.it

Nicola Fusco
Dipartimento di Matematica e Applicazioni
Università di Napoli
Via Cintia
I-80126 Napoli, Italy
E-mail address: n.fusco@unina.it

Progress in Nonlinear Differential Equations
and Their Applications, Vol. 51, 103–116
© 2002 Birkhäuser Verlag Basel/Switzerland

Interface Energies
and Structured Deformations in Plasticity

Gianpietro Del Piero

Abstract. I discuss the model of an elastic bar with interface energy at the jump points of the displacement function. If the interface energy is convex near the origin, the energy minima are attained by uniform limits of sequences of functions with a number of jumps growing to infinity, while their amplitudes decrease to zero. The limit elements are identified with mathematical objects called structured deformations. The model determines three regimes for the bar, which I call elastic, plastic, and fractured, and is able to describe the phenomenon of elastic unloading.

1. Introduction

The model of an elastic bar with bulk and interface energy has been successful in describing many aspects of material response observed in experiments. Specifically, a rather accurate description of fracture is obtained by assuming concave interface energies [7, 9, 16], while bi-modal energies reproduce the stress-strain behavior typical of damage [14, 15, 16].

Elastic-plastic behavior can be described in two different ways [9]. The first is to assume an interface energy convex in a neighborhood of the origin and concave anywhere else, and the second is to assume a periodic interface energy. Remarkably, both approaches are able to predict elastic unloading, a major difficulty encountered when constructing models for plasticity.

In this note I concentrate on the first model, adding some comments and explanations to the basic results presented in [9, 10, 16]. In Section 2 I give a characterization of the equilibrium configurations of the bar, and in Section 3 I state the minimum problem for the energy. In the same section, the minimum problem is decomposed in a way which easily leads to some general existence and non-existence results. In Section 4 I consider convex-concave interface energies, and for such energies I evaluate the global minimizers as functions of the elongation. I show that there is a range of elongations, depending on material constants and on the length of the bar, for which the energy minimum is not attained. This is precisely the *plastic regime* of the bar. In it, the convexity near the origin assumed for the interface energy produces minimizing sequences made of discontinuous functions, in which the number of jumps grows to infinity, the amplitudes decrease to zero, and the sum of the amplitudes remains bounded.

In Section 5 I show that the minimizing sequences admit a limit element in some generalized sense, and I analyze the nature of that limit. I show that it cannot be described, in general, by a single function. An adequate description requires two functions, representing the macroscopic displacement and the elastic part of the deformation, respectively. This pair of functions is called a *structured deformation*. Its ability to describe plastic behavior even outside the present one-dimensional context has been discussed elsewhere [12, 13, 18]. Finally, in Section 6 I show how elastic unloading can be taken into the picture. For this purpose, I state the incremental equilibrium problem from an equilibrium configuration represented by a structured deformation and I show that, as a consequence of energy minimization, the elastic part of the deformation stays constant when the elongation increases, and decreases when the elongation decreases.

2. Equilibrium of a bar with interface energy

Consider an elastic bar of length l, subject to axial displacements u. Let the displacements be allowed to have jump discontinuities, and with each jump associate an *interface energy*. In the absence of external loads, the total energy of the bar is

$$E(u) = \int_0^l w(u'(x))\, dx + \sum_{x \in S(u)} \theta([u](x)) . \qquad (2.1)$$

The functional E is defined over SBV$(0,l)$, the set of all functions u with bounded total variation over $(0,l)$ whose distributional derivative Du is a measure with null Cantor part [2]. Then Du is the sum of an absolutely continuous part u' and a jump part $[u]$. The jump part is concentrated on a finite or countable set $S(u)$, the *jump set* of u.

In (2.1), w is the *bulk energy density* and θ is the *interface energy*. They are non-negative functions defined over the real line and satisfying

$$w(0) = w'(0) = 0, \qquad\qquad \theta(0) = 0. \qquad (2.2)$$

Moreover, w is assumed to be strictly convex and to approach $+\infty$ at extreme deformations

$$\lim_{u' \to +\infty} w(u') = \lim_{u' \to -1} w(u') = +\infty, \qquad (2.3)$$

and θ is assumed to be lower semicontinuous.

Let us consider the *equilibrium configurations* of the bar subject to a prescribed elongation $\beta l > 0$

$$\int_0^l u'(x)\, dx + \sum_{x \in S(u)} [u](x) = \beta l \qquad (2.4)$$

and to the non-interpenetration constraint

$$[u](x) > 0 \qquad \forall x \in S(u). \tag{2.5}$$

In the following, β will be called the *strain* in the bar.

Denote by $\mathcal{A}_\beta$ the set of all functions in $\text{SBV}(0, l)$ which obey (2.4) and (2.5). In [16], the equilibrium configurations for the given β have been identified with the elements of the set

$$\mathcal{E}_\beta := \{ u \in \mathcal{A}_\beta \ \mid \ (u + \eta) \in \mathcal{A}_\beta \ \to \ \lim_{\lambda \to 0+} \tfrac{1}{\lambda}\left(E(u + \lambda\eta) - E(u) \right) \geq 0 \} \,. \tag{2.6}$$

Under the supplementary assumption that $\theta(0+) = 0$, it has been proved in [16] that all configurations in $\mathcal{E}_\beta$ satisfy the conditions

$$u'(x) = p, \quad \theta'([u](x)) = w'(p) = \sigma, \quad \sigma \leq \theta'(0+)\,, \tag{2.7}$$

and that all configurations in $\mathcal{A}_\beta$ which satisfy the same conditions with the inequality in $(2.7)_3$ replaced by strict inequality belong to $\mathcal{E}_\beta$. The constants p and σ are the *bulk deformation* and the *stress* in the bar, respectively. Notice that, because w strictly convex implies w' strictly increasing, the relation $(2.7)_2$ can be inverted, and from $(2.7)_3$ we have

$$p = (w')^{-1}(\sigma) \leq (w')^{-1}(\theta'(0+)) =: k\,. \tag{2.8}$$

3. Existence of global Minimizers

A sufficient condition for the existence of global minimizers is provided by Ambrosio's theorem of compactness and lower semicontinuity [1, 2, 4]. This theorem asserts that, if w is convex, if θ is concave, and if

$$\lim_{t \to +\infty} \tfrac{1}{t} w(t) = +\infty, \qquad \lim_{t \to 0+} \tfrac{1}{t} \theta(t) = +\infty, \tag{3.1}$$

then E is lower semicontinuous. Moreover, E is coercive by the assumptions (2.3) made on w, and it is known that a coercive lower semicontinuous functional has global minimizers [8].

Unfortunately, this existence condition does not apply to plasticity; indeed, it will be shown below that plastic-like response requires interface energies which are not concave and do not obey the limit condition $(3.1)_2$. In [16], the existence of global minimizers has been studied by adopting the following minimization technique. A global minimizer, if it exists, is an equilibrium configuration, and for such a configuration the expression (2.1) of the energy reduces to

$$E(u) = lw(p) + \sum_{k=1}^{N(u)} \theta([u](x_k))\,, \tag{3.2}$$

with $N(u)$ the number of the jump points x_k in $S(u)$. Moreover, the constraints (2.4), (2.5) take the form

$$[u](x_k) > 0, \qquad lp + \sum_{k=1}^{N(u)} [u](x_k) = l\beta. \tag{3.3}$$

Using $(3.3)_2$, we may eliminate p in (3.2). In this way, after setting $q_k := [u](x_k)$ we obtain the minimum problem for the functional

$$F^\beta(q) = l\, w(\, \beta - l^{-1} \sum_{k=1}^{\infty} q_k\,) + \sum_{k=1}^{\infty} \theta(q_k), \tag{3.4}$$

subject to the sole condition

$$q_k \geq 0. \tag{3.5}$$

In (3.4), q denotes the sequence $k \mapsto q_k$. By allowing the q_k to assume the value zero, displacements with a finite number of jumps can be represented by sequences with a finite number of non-zero elements.

The minimum problem for F^β can be split into two steps. The first step is to minimize F^β under the supplementary condition

$$\sum_{k=1}^{\infty} q_k = a \tag{3.6}$$

for a fixed $a > 0$. Under this condition, the first term in $F^\beta(q)$ is known, and to minimize F^β becomes equivalent to minimizing the sum of the $\theta(q_k)$. Let us set

$$\theta_s(a) := \inf \{\, \sum_{k=1}^{\infty} \theta(q_k) \mid q_k \geq 0, \sum_{k=1}^{\infty} q_k = a\, \}. \tag{3.7}$$

The function θ_s is the *subadditive envelope* of θ [16]. Thus, the first step in the minimization of F^β is the determination of the subadditive envelope of θ.

The second step is minimization with respect to a. It is made by minimizing the function

$$f^\beta(a) := l\, w(\beta - l^{-1}a) + \theta_s(a) \tag{3.8}$$

over all non-negative a. This function is coercive, because w is coercive by assumptions (2.3) and θ_s is non-negative by (3.7) and by the non-negativeness of θ. Moreover, w is continuous because it is convex, and therefore f^β is lower semicontinuous if θ_s is. Thus, the lower semicontinuity of θ_s implies the coerciveness and lower semicontinuity of f^β, and therefore the existence of global minimizers. A sufficient condition for the lower semicontinuity of θ_s is that θ be the sum of a Lipschitz continuous and of a piecewise constant, lower semicontinuous function [16].

The existence of minimizers for f^β is not sufficient to ensure the existence of minimizers for F^β. An existence theorem and a non-existence theorem are given in [16]. The existence theorem asserts that

Theorem 3.1 *If there is a $c > 0$ such that $\theta_s(a) = \theta(a)$ for all a in $[0, c]$, then there exist global minimizers for F^β in $\mathcal{A}_\beta$ for every $\beta > 0$,*

and the non-existence theorem says that

Theorem 3.2 *If there is a $c > l(1+k)$, with k defined by (2.8), such that $\theta_s(a) < \theta(a)$ for all a in $(0, c)$, then there is no minimizer for F^β in $\mathcal{A}_\beta$ for all β in the interval*

$$k < \beta < l^{-1}c - 1. \tag{3.9}$$

In view of the strict connection between non-existence of global minimizers and the occurrence of plastic deformation, I give here a sketch of the proof of the second theorem.

Sketch of proof of Theorem 3.2. Let q_β and a_β be global minimizers for F^β and f^β, respectively. It can be proved [16] that $N(q^\beta) = 0$ implies $\beta = p \leq k$ and that $N(q^\beta) > 0$ implies that every non-null element of q^β is greater than c, so that a_β is greater than c as well.

Now, by (3.3)$_2$ and (3.6),

$$p = \beta - l^{-1}a_\beta \tag{3.10}$$

and $p > -1$ by (2.2)$_2$. Then,

$$\beta > l^{-1}a_\beta - 1 > l^{-1}c - 1. \tag{3.11}$$

Consequently, if $c > l(1 + k)$ there is no minimizer a_β for β in the range (3.9).

4. Perfect plasticity

It is generally assumed that the interface energy is concave. In this case, θ_s coincides with θ and the infimum (3.7) is attained at sequences with only one non-zero element.

Concave interface energies are appropriate to describe the process of fracture, in accordance with the *cohesive crack* theory of Barenblatt [3]. To get plastic-like response, one needs interface energies which are convex in some neighborhood of the origin. An energy with this property is shown in Fig. 1. It is convex in $(0,d)$ and concave in $(d, +\infty)$. In the figure, c denotes the abscissa of the intersection of the graph of θ with the tangent at the origin:

$$\theta(c) = c\,\theta'(0+). \tag{4.1}$$

It can be seen [16] that the subadditive envelope of θ coincides with θ for $a > c$ and with the tangent to the curve (θ, a) at the origin for $a \leq c$:

$$\theta_s(a) = \begin{cases} a\,\theta'(0+) & \text{for } a < c, \\ \theta(a) & \text{for } a \geq c. \end{cases} \tag{4.2}$$

Let a_β be a global minimizer for the function f^β defined in (3.8). Then one of the three following alternatives is true.

(i) $a_\beta = 0$. In this case from (3.8) we have

$$f^\beta(a_\beta) \;=\; lw(\beta)\,, \tag{4.3}$$

and from (3.10) and (2.8) we get $p = \beta \le k$.

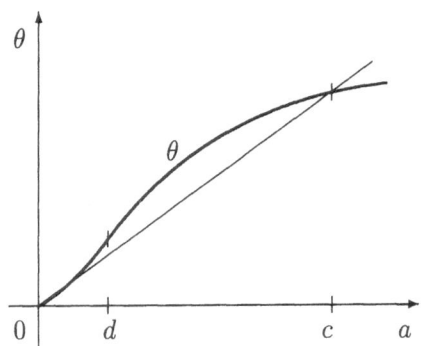

Figure 1.
Convex-concave interface energy.

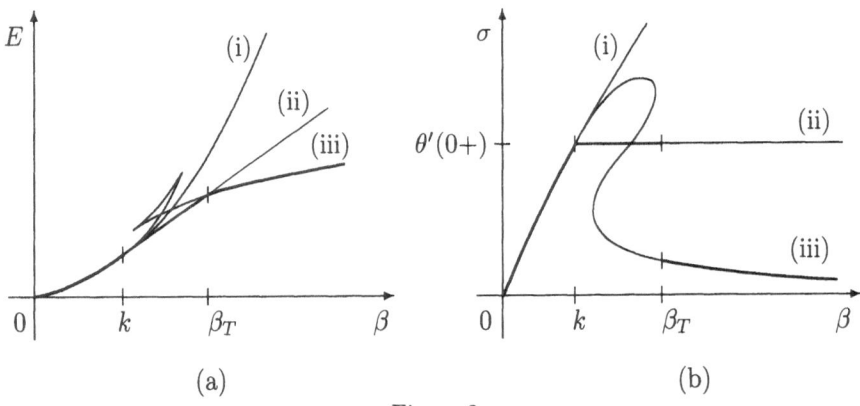

Figure 2.
Overall energy-strain (a) and stress-strain (b) curves.
(i) = elastic, (ii) = perfectly plastic, (iii) = fractured regime.
Bold lines = global minimizers.

(ii) $0 < a_\beta < c$. In this case by imposing that the derivative of f^β vanishes at a_β we have

$$w'(\beta - l^{-1}a_\beta) \;=\; \theta'_s(a_\beta)\,. \tag{4.4}$$

Then by (4.2)

$$\sigma = w'(\beta - l^{-1}a_\beta) = \theta'_s(a_\beta) = \theta'(0+) \,, \tag{4.5}$$

so that, recalling $(2.8)_3$,

$$p = \beta - l^{-1}a_\beta = (w')^{-1}(\theta'(0+)) = k \,. \tag{4.6}$$

The energy minimum is

$$f^\beta(a_\beta) = lw(k) + a_\beta\theta'(0+) = lw(k) + l(\beta - k)w'(k), \tag{4.7}$$

and from (4.6) we have $\beta = k + l^{-1}a_\beta$, so that

$$k < \beta < k + l^{-1}c \,. \tag{4.8}$$

(iii)   Finally, for $a_\beta \geq c$ we have

$$f^\beta(a_\beta) = lw(\beta - l^{-1}a_\beta) + \theta(a_\beta). \tag{4.9}$$

and $\beta > l^{-1}c - 1$ by (3.11).

If we assume $c > l(1 + k)$ as in the non-existence theorem, we may conclude that for $\beta \leq k$ there are only minimizers of type (i) and for $\beta < k + l^{-1}c$ there are only minimizers of type (iii), while for $k < \beta < k + l^{-1}c$ there is a competition between minimizers of types (ii) and (iii).

The expressions (4.2), (4.7) and (4.9) of the energy are plotted in Fig. 2.a. We see that there is a transition value β_T of β such that the global minimizers are of type (ii) for $k < \beta < \beta_T$ and of type (iii) for $\beta > \beta_T$. The stress-strain curve is shown in Fig. 2.b. It is determined by the equation

$$\sigma = \tfrac{1}{2} \tfrac{d}{d\beta} f^\beta(a_\beta) \,, \tag{4.10}$$

obtained by differentiating (3.8) with respect to β at $a = a_\beta$ and then using the equilibrium condition (4.4). We see that the curve consists of three branches, corresponding to the three types of minimizers seen above. The first branch is increasing, the second is constant, and the third is decreasing and close to zero. They characterize the *elastic*, the *perfectly plastic*, and the *fractured* regimes of the bar, respectively. The drop of σ at $\beta = \beta_T$ marks the occurring of fracture.

A peculiar property of the perfectly plastic regime is non-existence of global minimizers for F^β. Indeed, let a_β be a global minimizer for f^β of type (ii). Then for every sequence q with

$$q_k \geq 0, \quad \sum_{k=1}^{\infty} q_k = a_\beta \,, \tag{4.11}$$

from (3.4) and (4.6) we have

$$F^\beta(q) = lw(k) + \sum_{k=1}^{\infty} \theta(q_k). \tag{4.12}$$

Because a_β of type (ii) is smaller than c, each element q_k of q is smaller than c, and therefore $\theta(q_k) > q_k\theta'(0+)$, for θ as shown in Fig. 1. Then,

$$\sum_{k=1}^{\infty} \theta(q_k) > \theta'(0+) \sum_{k=1}^{\infty} q_k = \theta'(0+) a_\beta, \tag{4.13}$$

and from (4.12) and (4.7)$_1$ we conclude that

$$F^\beta(q) > f^\beta(a_\beta). \tag{4.14}$$

for all q obeying (4.11).

5. Structured deformations

In the preceding example, the infimum of E is not attained in $\mathcal{A}_\beta$ if $c > l(1+k)$ and $k < \beta < \beta_T$. Let us see whether it is attained outside this set. Consider the sequence $n \mapsto q^n$:

$$q_k^n = \begin{cases} n^{-1}l(\beta - k) & \text{for } k \leq n, \\ 0 & \text{for } k > n. \end{cases} \tag{5.1}$$

It is a minimizing sequence for F^β. Indeed,

$$\sum_{k=1}^{\infty} \theta(q_k^n) = n\,\theta(n^{-1}l(\beta - k)) = n\left(n^{-1}l(\beta - k)\,\theta'(0+) + o(n^{-1})\right), \tag{5.2}$$

and for $n \to \infty$ this sum converges to $l(\beta - k)\,\theta'(0+)$ which, by (4.7), determines the global minimum of f^β, and therefore the infimum of E. With each q^n we may associate the displacement

$$u_n(x) = kx + \varphi_n(x), \tag{5.3}$$

where φ_n are non-decreasing, piecewise constant functions with $\varphi_n(0) = 0$ and with n jumps of equal amount $n^{-1}l(\beta - k)$. Then each u_n belongs to $\mathcal{A}_\beta$, and $n \mapsto u_n$ is a minimizing sequence for E in $\mathcal{A}_\beta$.

Let φ^c be the Cantor function. If we take the subsequence $m \mapsto \varphi_m$, $m = 2^n - 1$, and if we select the locations x_k of the jumps in the following way

for $m = 1$: $x_k/l = 1/3$,
for $m = 2$: $x_k/l = 1/9,\ 3/9,\ 7/9$,
for $m = 3$: $x_k/l = 1/27,\ 3/27,\ 7/27,\ 9/27,\ 19/27,\ 21/27,\ 25/27$,

....

then $m \mapsto u_m$ converges uniformly to the function

$$u_\infty(x) := kx + l(\beta - k)\,\varphi^c(l^{-1}x). \tag{5.4}$$

Indeed, for every m and for every x in $[0, l]$ we have

$$|u_m(x) - u_\infty(x)| \leq m^{-1} l(\beta - k) . \tag{5.5}$$

In $m \mapsto u_m$, the absolutely continuous parts of the distributional derivatives Du_m satisfy $u'_m(x) = k$ for almost every x; therefore, they coincide with the absolutely continuous part u'_∞ of Du_∞. The singular parts of Du_m have the same total variation $l(\beta - k)$ as Du_∞, but they are jump singularities, while the singular part of Du_∞ is a diffuse measure of the Cantor type. Thus, u_∞ does not belong to SBV$(0,l)$, and therefore to $\mathcal{A}_\beta$.

Plastic deformations with slips distributions reminiscent of the Cantor function can be seen in experiments, see e.g. Kleiser and Bocek [17], quoted in [6]. Nevertheless, uniformly distributed slips are more frequently observed, at least at the macroscopic level. Within the model discussed here, a macroscopically uniform distribution of jumps can be obtained by subdividing the bar into a sufficiently large number h of intervals of length lh^{-1} and by locating the jumps, in each interval, with the same strategy applied above to the interval $(0,l)$. When the number of jumps tends to infinity, we obtain functions u^h of the type

$$u^h(x) = kx + l(\beta - k)h^{-1}(r - 1 + \varphi^c(xhl^{-1} - r + 1))$$
$$\text{for } \tfrac{r-1}{h}l < x < \tfrac{r}{h}l, \qquad r \in \{1, 2, \ldots h\}. \tag{5.6}$$

Consider the sequence $h \mapsto u^h$. It converges uniformly to

$$u(x) := kx + l(\beta - k)xl^{-1} = \beta x. \tag{5.7}$$

However, the function u does not describe adequately the limit of $h \mapsto u^h$. Indeed, each u^h has a derivative with a Cantor part with total variation $l(\beta - k)$, while no Cantor part is present in u. That some singular measure has been lost in the limit procedure is revealed by the inequality

$$\lim_{h \to \infty} (u^h)'(x) \neq \left(\lim_{h \to \infty} u^h(x) \right)' . \tag{5.8}$$

It is explained in [12, 13] that complete information about the limit of $h \mapsto u^h$ is provided by the pair (u, v), where u and v are the uniform limits of $h \mapsto u^h$ and $h \mapsto (u^h)'$, respectively. This pair is called a *structured deformation*. It has been proved in [11] that the energy of a one-dimensional structured deformation is

$$E(u, v) = \int_0^l (w_c(v(x)) + \theta'(0+)(u'(x) - v(x))) \, dx + \sum_{x \in S(u)} \theta_s([u](x)) , \tag{5.9}$$

where w_c is the convex envelope of w and θ_s is the subadditive envelope of θ. In the case considered here, we have $u(x) = \beta x$ and $v(x) = k$. Moreover, $S(u)$ is the empty set and $w_c = w$ because w is convex. Then $E(u, v)$ is equal to $f^\beta(a_\beta)$ as given by (4.7), and therefore the global energy minimum is attained at (u, v).

6. Elastic unloading

One of the main difficulties encountered in modelling plastic behavior is to describe correctly the phenomenon of elastic unloading. Let us assume that, for increasing β, the bar follows the curve of the global energy minima of Fig. 2.a. Then it seems reasonable to assume that the same occurs when β decreases. But experiments show a completely different behavior: when β decreases starting from a point on the curve (ii) (plastic regime), the bar follows a line parallel to the tangent at the origin to the curve (i) (elastic regime).

If we identify points on the curve (ii) with structured deformations (u, v), we may decompose the total deformation u' into the sum of an *elastic part* v and a *plastic part* $u' - v$. Then at elastic unloading the plastic part stays constant, and the whole decrease of β is taken by the elastic part.

This scheme is not new: most of the engineering theories of plasticity are based on additive or multiplicative decompositions of the deformation gradient. What is new in the present analysis is the possibility of *predicting* elastic unloading as a result of energy minimization. The idea is to minimize the increment of energy occurring when the strain changes from β to $\beta + \partial\beta$ [9]. Let the bar be at equilibrium in a configuration described by the structured deformation (u, v), and let η be a perturbation in SBV$(0,l)$, which changes the strain from β to $\beta + \partial\beta$:

$$\int_0^l \eta'(x)\, dx \; + \sum_{x \in S(\eta)} [\eta](x) \;=\; l\, \partial\beta. \tag{6.1}$$

Then by (5.9) the energy of the structured deformation $(u + \eta, v + \eta')$ is

$$E(u + \eta, v + \eta')$$

$$= \int_0^l (w_c(v(x) + \eta'(x)) + \theta'(0+)(u'(x) - v(x)))\, dx + \sum_{x \in S(u+\eta)} \theta_s([u + \eta](x))$$

$$= E(u + v) + \int_0^l (w_c(v(x) + \eta'(x)) - (w_c(v(x))))\, dx$$

$$+ \sum_{x \in S(u+\eta)} \theta_s([u + \eta](x)) - \sum_{x \in S(u)} \theta_s([u](x)). \tag{6.2}$$

Returning to the situation examined in Sect. 5, for an equilibrium configuration in the plastic regime we have that $u(x) = \beta x, v(x) = k$, $S(u)$ is empty, and $w_c = w$. Moreover, for sufficiently small $\partial\beta$ we may set $\theta_s = \theta$. Indeed, in the case of empty $S(u)$ the jump terms in (6.2) reduce to $\theta_s([\eta](x))$, and $[\eta](x)$ is smaller than $l\, \partial\beta$ by (6.1). But, by (4.2), $\theta_s(a) = \theta(a)$ for sufficiently small a. Then we have the problem of minimizing the functional

$$\Phi(\eta) := \int_0^l (w(k + \eta'(x)) - w(k))\, dx \; + \sum_{x \in S(\eta)} \theta([\eta](x)) \tag{6.3}$$

in the class of all η in SBV$(0,l)$ which satisfy the constraint (6.1) and the non-interpenetration condition

$$[\eta](x) > 0 \qquad \forall x \in S(\eta).\tag{6.4}$$

Let us evaluate two lower bounds for Φ.

(i) If $S(\eta) = \emptyset$, then from Jensen's inequality and (6.1)

$$\Phi(\eta) \geq l(w(k + l^{-1} \int_0^l \eta'(x)\,dx) - w(k)) = l(w(k + \partial\beta) - w(k)).\tag{6.5}$$

(ii) If $S(\eta) \neq \emptyset$, recalling that θ is convex in a neighborhood of the origin and that $w'(k) = \theta'(0+)$ by (2.8),

$$\Phi(\eta) \geq w'(k) \int_0^l \eta'(x)\,dx + \theta'(0+) \sum_{x \in S(\eta)} [\eta](x) = l\theta'(0+)\partial\beta.\tag{6.6}$$

Now fix λ in $[0,1]$ and consider a sequence $n \mapsto \eta_n$ such that

$$\eta_n'(x) = \lambda\partial\beta, \qquad \sum_{x \in S(\eta_n)} [\eta_n](x) = (1 - \lambda)l\partial\beta.\tag{6.7}$$

For instance, we may assume that η_n has n jumps, each of amount $n^{-1}(1-\lambda)l\partial\beta$. Then the condition (6.4) requires

$$(1 - \lambda)\,\partial\beta \geq 0.\tag{6.8}$$

Let us assume that all configurations $u + \eta_n$ are equilibrated. Then the condition (2.8) requires $u'(x) + \eta_n'(x) \leq k$, and because $u'(x) = k$ we have $\eta_n'(x) \leq 0$, and therefore

$$\lambda\,\partial\beta \leq 0.\tag{6.9}$$

If $\partial\beta$ is positive, then $\lambda = 0$ and $\eta_n' = 0$. Then

$$\Phi(\eta_n) = n\theta(n^{-1}l\partial\beta) = n(\theta'(0+)n^{-1}l\partial\beta + o(n^{-1})),\tag{6.10}$$

and for $n \to \infty$ we have that $n \mapsto \Phi(\eta_n)$ converges to the lower bound (6.6). If $\partial\beta$ is negative, then $\lambda = 1$ by (6.8), so that

$$\Phi(\eta_n) = l(w(k + \partial\beta) - w(k)),\tag{6.11}$$

and the energy converges to the lower bound (6.5).

In both cases, $n \mapsto \eta_n$ is a minimizing sequence. For $\partial\beta > 0$ we have $\eta_n'(x) = 0$, and $n \mapsto \eta_n$ converges uniformly to $\eta_\infty(x) := x\,\partial\beta$. Then the minimum of Φ is attained at the structured deformation $(u + \eta_\infty, v)$. For $\partial\beta < 0$, we have $\eta_n(x) = x\,\partial\beta = \eta_\infty(x)$, $\eta_n'(x) = \partial\beta$, and the minimum is attained at $(u + \eta_\infty, v + \eta_\infty')$.

In the first case η_∞ is a purely plastic incremental deformation in which, with increasing $\partial\beta$, the elastic part stays constant at the value k and, by (2.7)$_2$, the stress stays constant at the value $\theta'(0+)$. In the second case η_∞ is purely elastic;

since there are no jumps in η_∞, the plastic part of the incremental deformation stays constant at $u' - v$, while the elastic part varies as $\beta + \partial\beta$, and the stress as $w(\beta + \partial\beta)$. These two responses are shown in Fig. 3. They correspond to *plastic loading* and to *elastic unloading*, respectively.

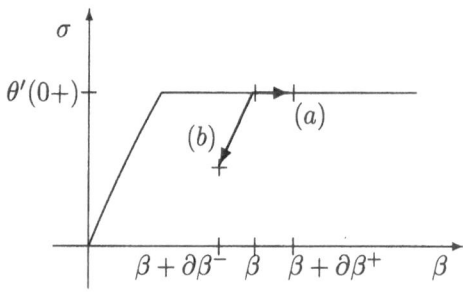

Figure 3.
Plastic loading (a) and elastic unloading (b).

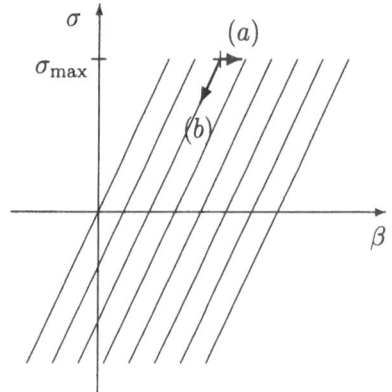

Figure 4.
Plastic loading (a) and elastic unloading (b) in periodic plasticity.

7. Periodic plasticity

An alternative model for plasticity, which does not require the use of structured deformations, is obtained by taking a periodic interface energy θ. As shown in Fig. 4, a periodic θ generates a family of parallel equilibrium curves in the stress-strain plane. Plastic deformation is identified with jumps from one equilibrium curve to the other, and elastic unloading results just from following an equilibrium curve as far as possible when β decreases. This model has the advantage of predicting yielding in compression as well as in tension. In this sense, it fits the *kinematical hardening* model for the plasticity of metals in the special case of perfect plasticity. Indeed, there is actually no hardening, because the maximum stress σ_{max} does not change when jumping from one equilibrium curve to the other. For further details see [7, 9].

Acknowledgement. This research has been supported by the Programma Nazionale Cofinanziato 2000 *Modelli Matematici per la Scienza dei Materiali.*

References

[1] L. Ambrosio, *A compactness theorem for a new class of functions of bounded variation,* Boll. Un. Mat. Ital., **3-B** (1989), 857–881.

[2] L. Ambrosio, *Free discontinuity problems and special functions with bounded variation,* Progress in mathematics, **168** (1998), 15–35.

[3] G.I. Barenblatt, *The mathematical theory of equilibrium cracks in brittle fracture,* Adv. Appl. Mech., **7** (1962), 55–129.

[4] A. Braides, *Approximation of free-discontinuity problems,* Springer Lecture Notes in Mathematics **1694** (1998), Springer-Verlag, Berlin.

[5] A. Braides and A. Coscia, *A singular perturbation approach to variational problems in fracture mechanics,* Math. Mod. Meth. Appl. Sci, **3** (1993), 303–340.

[6] A. Carpinteri, B. Chiaia and P. Cornetti, *Static-kinematic duality and the principle of virtual work for fractal media,* in: Proc. ECCOMAS 2000, European Congress on Computational Methods in Applied Sciences and Engineering (2000)

[7] R. Choksi, G. Del Piero, I. Fonseca and D.R. Owen, *Structured deformations as energy minimizers in models of fracture and hysteresis,* Math. and Mech. of Solids, **4** (1999), 321–356.

[8] B. Dacorogna, *Direct methods in the calculus of variations,* (1989), Springer-Verlag, Berlin.

[9] G. Del Piero, *Towards a unified approach to fracture, yielding, and damage,* in: E. Inan and K.Z. Markov, Eds., Continuum Models and Discrete Systems, (World Scientific, Singapore) (1998), 679–692.

[10] G. Del Piero, *One-dimensional ductile-brittle transition, yielding, and structured deformations,* in: P. Argoul, M. Frémond M. and Q.S. Nguyen, Eds., Variations of domains and free boundary problems in solid mechanics, (Kluwer, Dordrecht) (1999), 203–210.

[11] G. Del Piero, *The energy of a one-dimensional structured deformation,* Math. and Mech. of Solids, **6** (2001), 387–408.

[12] G. Del Piero and D.R. Owen, *Structured deformations of continua,* Arch. Rational Mech. Analysis, **124** (1993), 99–155.

[13] G. Del Piero and D.R. Owen, *Structured deformations,* Quaderni dell'Istituto Nazionale di Alta Matematica, (2000), Centro Stampa 2P, Firenze.

[14] G. Del Piero and L. Truskinovsky, *A one-dimensional model for localized and distributed failure,* J. Phys. IV France **8** (1998), 95–102.

[15] G. Del Piero and L. Truskinovsky, *Macro and micro cracking in one-dimensional elasticity,* Int. J. of Solids and Structures **38** (2001), 1135–1148.

[16] G. Del Piero and L. Truskinovsky, *Elastic bars with cohesive energy,* in preparation.

[17] T. Kleiser and M. Bocek, *The fractal nature of slip in crystals,* Zeits. Metal. **77** (1986), 582–587.

[18] D.R. Owen, *Refined continuum models of plasticity and related phenomena,* in: A. Khan, Ed., Proceedings of the Seventh International Symposium on Plasticity and its Current Applications, Neat Press, Fulton, MD, (1999).

Gianpietro Del Piero
Dipartimento di Ingegneria
Università di Ferrara
via Saragat 1
I-44100 Ferrara
Italia
E-mail address: gdpiero@ing.unife.it

Progress in Nonlinear Differential Equations
and Their Applications, Vol. 51, 117–140

Higher Order Variational Problems and Phase Transitions in Nonlinear Elasticity

Irene Fonseca and Giovanni Leoni

Abstract. Higher order variational problems ask for new techniques which do not reduce to mere generalizations of their counterparts for first order problems. These challenges will be illustrated within the context of phase transitions for nonlinear elastic materials, and of higher order lower semicontinuity results recently obtained.

1. Introduction

Higher order variational problems appear often in the engineering literature – in connection with the so-called gradient theories of phase transitions within elasticity regimes, and where higher order terms give rise to surface energies; in the study of equilibria of micromagnetic materials where mastery of second order energies (here accounting for the exchange energy) is required (see [57], [101]; see also [33], [42], [48], [49], [67], [87], [88], [89], [120]); in the Blake-Zisserman model for image segmentation in computer vision (see [38], [39], [40]; see also [54]), which seats squarely among second-order free discontinuity models that may be recast as higher order Griffiths' models for fracture mechanics (see [9], [26], [30], [64], [76], [77], [78], [79]); in several mechanics of thin films, delamination and blistering, etc., with bending energy appearing as the higher order term; in the analysis of many other Landau theories with non-convex variational problems regularized by higher order singular perturbations; etc.

Energy functionals may include lower dimensional order terms to take into account interfacial energies and discontinuities of underlying fields (see [12]), although in what follows we will neglect the role played by these terms and we will focus on the added difficulties inherent to the presence of derivatives of order two or more.

The main difficulty encountered with the handling of the higher order energies is that one would be tempted to treat them as first order problems. However, a lesson to be learned is that techniques for higher order variational problems do not reduce to mere generalizations of their counterparts for first order problems (e.g., by assuming k-quasiconvexity in place of quasiconvexity, see [46], [97], [100]). Indeed, although functionals depending *uniquely* on the highest order derivatives can be treated easily, those where lower order terms are present require new ideas and new tools to handle the localization and truncation of lower order terms.

To illustrate, consider an energy functional

$$I(u) := \int_\Omega f(x, u, \nabla u, \ldots, \nabla^k u)\, dx \tag{1}$$

where $\Omega \subset \mathbb{R}^N$ is an open, bounded domain, $u : \Omega \to \mathbb{R}^d$, $N, d \geq 1$, $u := (u_1, \ldots, u_d)$, $\nabla u \in \mathbb{R}^{d \times N}$, and $(\nabla u)_{ij} := \frac{\partial u_i}{\partial x_j}$ for $i \in \{1, \ldots, d\}$, $j \in \{1, \ldots, N\}$.

What assumptions on f guarantee that if $\{u_n\}$ is a sequence bounded in $W^{k,1}(\Omega; \mathbb{R}^d)$ and if $u_n \to u$ in $W^{k-1}(\Omega; \mathbb{R}^d)$ then

$$I(u) \leq \liminf_{n \to +\infty} I(u_n)?$$

The usual technique amounts to keep unchanged the higher order, oscillating terms, while freezing the lower order, strongly converging terms, by means of suitable truncations and the use of Scorza-Dragoni-type uniform continuity theorems. Truncating gradients so that they remain gradients may be achieved through the techniques of maximal functions and of Fourier multipliers (see [1], [109], [110]) in those cases where the bulk energy density f has superlinear growth (see the proof of Lemma 2.15 in [74]). In fact, the success of this approach relies heavily on p-equi-integrability, and thus cannot be extended to the case $p = 1$ where one replaces weak convergence in $W^{k,1}(\Omega; \mathbb{R}^n)$ with the natural convergence, i.e., strong convergence in $W^{k-1,1}(\Omega; \mathbb{R}^n)$ and bounds on the L^1 norms of the k-th order derivatives. As it turns out, when f grows at most linearly at infinity many seemingly simple questions, long ago answered within the realm of first order problems, still defy all attempts when we deal with order two or more. As an example, a standing open problem is the following (in the case where $k = 1$ this question was first answered by Fonseca and Müller in [72] and [73] and later improved by Fonseca and Leoni in [65] Theorem 1.8):

is it true that if $f : \Omega \times \mathbb{R}^s \times \mathbb{R}^{d \times N^k} \to [0, \infty)$ is a Borel integrand, with $s := d + d \times N + \ldots + d \times N^{k-1}$, if $f(x, \mathbf{v}, \cdot)$ is k-quasiconvex, with $k \geq 2$ and $\mathbf{v} := (u, \nabla u, \ldots, \nabla^{k-1} u)$, if f satisfies "reasonable" continuity properties with respect to x and $\mathbf{v}$, if

$$C|\xi| - \frac{1}{C} \leq f(x, \mathbf{v}, \xi) \leq C(1 + |\xi|)$$

for all $(x, \mathbf{v}) \in \Omega \times \mathbb{R}^s$, if $u \in W^{k-1,1}(\Omega; \mathbb{R}^d)$, $\nabla^k u \in BV(\Omega; \mathbb{R}^{d \times N^k})$, and if $\{u_n\}$ is a sequence of functions in $W^{k,1}(\Omega; \mathbb{R}^d)$ converging to u in $W^{k-1,1}(\Omega; \mathbb{R}^d)$, then

$$\int_\Omega f(x, u, \ldots, \nabla^k u)\, dx \leq \liminf_{n \to \infty} \int_\Omega f(x, u_n, \ldots, \nabla^k u_n)\, dx?$$

2. Phase transitions

The asymptotic behavior of functionals of the type

$$J_\varepsilon(v; \Omega) := \int_\Omega \frac{1}{\varepsilon} W(v) + \varepsilon |\nabla v|^2 \, dx \tag{2}$$

has been extensively studied within the context of phase transitions. Adopting a Gibbs' criterium for equilibria, the energy of a stable fluid in a 2-phase mode may be identified with minima of the problem

$$(P_0) \qquad \text{minimize} \quad \int_\Omega W(u) \, dx \qquad \text{with} \quad \int_\Omega u \, dx = m$$

where Ω is an open, bounded set in $\mathbb{R}^N$ and W is a non-negative bulk energy density with $\{W = 0\} = \{a, b\}$, $a, b \in \mathbb{R}$, $a < b$. Clearly, if $m < a$ or $m > b$ then the minimizer of (P_0) is simply $u \equiv m$ and no transitions are energetically favorable. However, if $m \in (a, b)$ then this problem admits infinitely many solutions, those of the form

$$u(x) := \begin{cases} a & \text{if } x \in D, \\ b & \text{if } x \in \Omega \setminus D, \end{cases}$$

where D is any measurable subset of Ω with measure $\frac{\mathcal{L}^N(D)}{\mathcal{L}^N(\Omega)} = \frac{b-m}{b-a} =: \theta$. This lack of uniqueness is due to the fact that interfaces are allowed to form without a concomitant increase of energy.

In 1893 Van der Waals proposed a gradient theory as a selection criteria for physically preferred solutions which takes into account interfacial energy. Important developments of this theory concerning the interfacial energy between phases were later obtained by Cahn and Hilliard [35]. Upon rescaling, the new penalized energy now reads as (2), and the minimization problem becomes

$$(P_\varepsilon) \qquad \text{minimize} \quad J_\varepsilon(u; \Omega) \qquad \text{with} \quad \int_\Omega u \, dx = m.$$

In 1983 Gurtin [85] conjectured that minimizers for (P_ε) converge to minimizers of (P_0) with minimal interfacial energy. This conjecture was proved in 1984 in the scalar case (where $N = 1$) by Carr, Gurtin and Slemrod [37]. Precisely, if $\Omega = (0, 1)$ then (P_ε) admits only two minimizers ($u_\varepsilon(x)$ and $u_\varepsilon(1 - x)$), and moreover

$$J_\varepsilon(u_\varepsilon) = \varepsilon K_0 + O\left(e^{-C/\varepsilon}\right)$$

where $K_0 := 2 \int_a^b \sqrt{W(s)} \, ds$ stands for the surface energy per unit area, and C is a positive constant.

In the higher dimensional case ($N > 1$) Gurtin's conjecture was proved independently by Modica [98] and by Sternberg [111]. The approach in both [98] and [111] uses De Giorgi's notion of Γ-convergence ([53]; see also [4], [29], [50]) and follows closely ideas of Modica and Mortola [99] who studied a similar functional proposed by De Giorgi in a completely different physical context. In particular, the result of Modica [98] only holds for minimizers of (P_ε), as it relies on results on the nature of minimizers of (P_0) obtained by Gonzalez, Massari and Tamanini [83], but it requires no regularity hypotheses on W beyond continuity, while Sternberg [111], under stronger regularity hypotheses on W, proved that

$$\Gamma - \lim_{\varepsilon \to 0^+} J_\varepsilon(u_0; \Omega) = \begin{cases} K_0 \text{Per}_\Omega(E) & \text{if} \quad u = \chi_E a + (1 - \chi_E) b, \ |E| = \theta \, |\Omega|, \\ & \qquad u \in BV(\Omega; \{a, b\}), \\ +\infty & \text{otherwise.} \end{cases}$$

(3)

The theory of Γ-convergence guarantees, in particular, that preferred designs are those which exhibit minimal interfacial area for the given volume fraction θ.

Generalizations of (2)–(3) were obtained by Bouchitté [25] and by Owen and Sternberg [104] for the undecoupled problem, in which the integrand in J_ε has the form $\varepsilon^{-1} f(x, v(x), \varepsilon \nabla v(x))$. For the study of local minimizers we refer to Kohn and Sternberg [92].

The vector-valued setting, where $u : \Omega \to \mathbb{R}^d$, $\Omega \subset \mathbb{R}^N$, $d, N > 1$, was considered in [20], [75] and [111], where K_0 is replaced by

$$K_1 := \inf \Big\{ \int_{-L}^{L} W(g(s)) + |g'(s)|^2 \, ds : L > 0, \ g \text{ piecewise } C^1, \ g(-L) = a,$$

$$g(L) = b \Big\}.$$

The case where W has more than two wells was addressed by Baldo [16] (see also Sternberg [112]), and later generalized by Ambrosio [7].

Motivated by questions within the realm of elastic solid-to-solid phase transitions (see [19], [43], [91]) we now consider the corresponding problem for gradient vector fields where $u : \Omega \to \mathbb{R}^d$ stands for the deformation, and in place of J_ε we introduce

$$I_\varepsilon(u; \Omega) := \begin{cases} \displaystyle\int_\Omega \frac{1}{\varepsilon} W(\nabla u) + \varepsilon |\nabla^2 u|^2 \, dx & \text{if } u \in W^{2,2}(\Omega; \mathbb{R}^d), \\ +\infty & \text{otherwise.} \end{cases}$$

(5)

The analysis of this model has defied considerable mathematical efforts during the past decade, and a significant contribution for two distinct classes of constitutive assumptions was made recently by Conti, Fonseca and Leoni [44].

An intermediate case between (2) and (5), where the non-convex potential depends on u and the singular perturbation on $\nabla^2 u$, has been recently studied by Fonseca and Mantegazza [71] (for other generalizations see [63]). Also, in the two-dimensional case and when W vanishes on the unit circle (5) reduces to the so-called Eikonal functional which arises in the study of liquid crystals (see [13]) as well as in blistering of delaminated thin films (see [103]). Recently, the Eikonal problem has received considerable mathematical attention, but in spite of substantial partial progress (see [11], [14], [59], [90]) its Γ-limit remains to be identified.

Going back to the results obtained in [44] concerning (5), we first notice that frame-indifference requires that $W(\xi) = W(R\xi)$ for all ξ and all $R \in SO(N)$, where $SO(N)$ is the set of rotations in $\mathbb{R}^N$. Therefore, and by analogy with the hypotheses initially placed on (2), if we assume that $W(A) = 0 = W(B)$ then $\{W = 0\} \subset SO(N)A \cup SO(N)B$. Also, in order to guarantee the existence of "classical" (as opposed to measure-valued) non-affine solutions for the limiting problem, and in view of Hadamard's compatibility condition for layered deformations (see also Ball and James [19]), the two wells must be rank-one connected. Hence, so as to be able to construct gradients taking values only on $\{A, B\}$ and layered perpendicularly to ν, we assume that

$$A - B = a \otimes \nu$$

for some $a \in \mathbb{R}^N$ and $\nu \in S^{N-1} := \partial B(0, 1) \subset \mathbb{R}^N$, and we simplify (greatly!) the problem by removing the frame-indifference constraint, and setting simply

$$\{W = 0\} = \{A, B\}.$$

Without loss of generality we may assume that

$$A = -B = a \otimes e_N.$$

Since now interfaces of minimizers must be planar with normal e_N (see [19]), at first glance the analysis may seem to be significantly less involved as compared with the initial problem (2) which requires the handling of minimal surfaces. However, it turns out that the PDE constraint curl $= 0$ imposed on the admissible fields presents numerous difficulties to the characterization of the Γ-limsup. In particular, if, say, ∇u has a layered structure with two interfaces then it is possible to construct a "realizing" (effective, or recovering) sequence nearby each interface, but the task of gluing together the two sequences on a suitable low-energy intermediate layer is very delicate. In order to illustrate the difficulties encountered here, we explain briefly how we would "normally" undertake the heuristic argument to glue together two optimal sequences $\{(u_n, \varepsilon_n)\}$, corresponding to an interface of a cylindrical body Ω at a given height h, and $\{(v_n, \delta_n)\}$, corresponding to an interface at a height h', $h' > h$.

First we must convince ourselves that the sequences $\{\varepsilon_n\}$ and $\{\delta_n\}$ (related to the "periodicity" of the ripples of the optimal fine structure near each interface) may be taken to be the same. This is by no means trivial (although true!). But let us take this for granted, and, as it is usual, we consider as a candidate for the two-interface situation a convex combination

$$w_{k,n} := \varphi_k u_n + (1 - \varphi_k) v_n$$

where φ_k is a smooth cut-off function, with $\{0 < \varphi_k < 1\} \subset L_{k,n}$ and $L_{k,n}$ is a horizontal layer intermediate between heights h and h'. The crux of the problem is to choose $L_{k,n}$ in a judicious way so that no extra energy is added to the system by the new sequence $\{w_{k,n}\}$.

Using De Giorgi's Slicing Method, we slice horizontally the layer between heights h and h' into M horizontal sub-layers L_k of width $(h' - h)/M$. In view of the fact that $||\nabla\varphi_k||_\infty = O(M)$, we then have

$$\sum_{k=1}^{M} \int_{L_k} \frac{1}{\varepsilon_n} W(w_{k,n}) + \varepsilon_n |\nabla w_{k,n}|^2 \leq O\left(\frac{1}{\varepsilon_n}\right) + \varepsilon_n M^2 ||u_n - v_n||^2. \tag{6}$$

Choosing $k = k(n)$ such that

$$\int_{L_{k(n)}} \frac{1}{\varepsilon_n} W(w_{k(n),n}) + \varepsilon_n |\nabla w_{k(n),n}|^2 \leq \frac{1}{M} \sum_{k=1}^{M} \int_{L_k} \frac{1}{\varepsilon_n} W(w_{k,n}) + \varepsilon_n |\nabla w_{k,n}|^2, \tag{7}$$

it is clear that by setting $M = O\left(\frac{1}{\varepsilon_n \sqrt{||u_n - v_n||_{L^2}}}\right)$, and using the fact that the admissible sequences $\{u_n\}$ and $\{v_n\}$ satisfy $||u_n - v_n||_{L^2} \to 0$ in the intermediate layer between heights h and h', we may conclude that

$$\lim_{n\to\infty} \int_{L_{k(n)}} \frac{1}{\varepsilon_n} W(w_{k(n),n}) + \varepsilon_n |\nabla w_{k(n),n}|^2 = 0.$$

Suppose now that we want to extend this argument to the present setting involving second order derivatives. Since $||\nabla^2 \varphi_k||_\infty = O(M^2)$, the estimate (6) now becomes

$$\sum_{k=1}^{M} \int_{L_k} \frac{1}{\varepsilon_n} W(\nabla w_{k,n}) + \varepsilon_n |\nabla^2 w_{k,n}|^2 \leq O\left(\frac{1}{\varepsilon_n}\right) + \varepsilon_n M^4 ||u_n - v_n||^2,$$

and, seeking *equi-partition of energy* as we have done above, we are led to

$$\varepsilon_n M^4 ||u_n - v_n||^2 = O\left(\frac{1}{\varepsilon_n}\right).$$

This would entail $M = O\left(\frac{1}{\sqrt{\varepsilon_n}\sqrt{||u_n - v_n||_{L^2}}}\right)$, and thus the upper bound in (7) would be

$$\int_{L_{k(n)}} \frac{1}{\varepsilon_n} W(\nabla w_{k,n}) + \varepsilon_n |\nabla^2 w_{k,n}|^2 \leq \frac{1}{M} O\left(\frac{1}{\varepsilon_n}\right) = O\left(\sqrt{\frac{||u_n - v_n||_{L^2}}{\varepsilon_n}}\right).$$

Therefore, in order to ensure that the extra energy in the layer $L_{k(n)}$ does not affect the optimality of the sequence, we would have to guarantee that $\{u_n - v_n\}$ goes to zero in L^2 faster than ε_n, and whether or not this holds it remains an open question!

The restrictive constitutive hypotheses placed on W in Theorems 2.1 and 2.2 below will allow us to find alternative ways in which the gluing is successful. We must, however, accept the fact that matching in one single swift step works well for first order problems but it is simply too abrupt when dealing with higher order derivatives. Indeed, the proofs of Theorems 2.1 and 2.2 are strongly hinged on a two-step matching technique where the control of Poincaré and Poincaré-Freidrichs' constants is carefully kept.

Theorem 2.1 ([44], Theorem 1.3). *Let $\Omega \subset \mathbb{R}^N$ be an open, bounded, simply connected domain with Lipschitz boundary. Assume that W satisfies the conditions*

(H_1) *W is continuous, $W(\xi) = 0$ if and only if $\xi \in \{A, B\}$, where $A = -B = a \otimes e_N$ for some $a \in \mathbb{R}^d \backslash \{0\}$;*

(H_2) *there exists $C > 0$ such that*

$$W(\xi) \geq C|\xi| - \frac{1}{C}$$

for all $\xi \in \mathbb{R}^{d \times N}$;

(H_3) *$W(\xi) \geq W(0, \xi_N)$ where $\xi = (\xi', \xi_N) \in \mathbb{R}^{d \times (N-1)} \times \mathbb{R}^d$.*

Suppose, in addition, that W is differentiable at A and B. Let $u \in W^{1,1}(\Omega; \mathbb{R}^d)$, with $\nabla u \in BV(\Omega; \{A, B\})$. Then there exists $K > 0$ such that

$$\Gamma - \lim_{\varepsilon \to 0^+} I_\varepsilon(u; \Omega) = K \operatorname{Per}_\Omega(E),$$

where $\nabla u = (1 - \chi_E(x)) A + \chi_E(x) B$ for $\mathcal{L}^N$ a.e. $x \in \Omega$.

The hypothesis (H_3) entails a one dimensional (geodesic) character to the asymptotic problem. Indeed, it can be shown that K reduces to the analog of the constant K_1 introduced in (4), precisely,

$$K := \inf \left\{ \int_{-L}^{L} W(0, g(s)) + |g'(s)|^2 \, ds \; : \; L > 0, \; g \text{ piecewise } C^1, \right.$$

$$\left. g(-L) = -a, \; g(L) = a \right\}.$$

Theorem 2.2 ([44], Theorem 1.4). *Let $\Omega \subset \mathbb{R}^N$ be an open, bounded, simply connected domain with Lipschitz boundary. Assume that W satisfies the conditions (H_1),*

(H_4) *W is even in each variable ξ_i, $i = 1, \cdots, N - 1$, that is $W(\xi_1, \cdots, -\xi_i, \cdots, \xi_N) = W(\xi_1, \cdots, \xi_i, \cdots, \xi_N)$ for each $i = 1, \cdots, N - 1$, and that there exist an exponent $p \geq 2$, constants $c, C, \rho > 0$ and a convex function $g : [0, \infty) \to [0, \infty)$, with $g(s) = 0$ if and only if $s = 0$, such that g is derivable in $s = 0$, it satisfies the doubling condition $g(2t) \leq cg(t)$ for all $0 \leq t \leq \rho$,*

$$g(|\xi - A|) \leq W(\xi) \leq cg(|\xi - A|) \quad \text{if } |\xi - A| \leq \rho,$$
$$g(|\xi - B|) \leq W(\xi) \leq cg(|\xi - B|) \quad \text{if } |\xi - B| \leq \rho,$$

and

$$\frac{1}{C} |\xi|^p - C \leq W(\xi) \leq C(|\xi|^p + 1)$$

for all $\xi \in \mathbb{R}^{d \times N}$. Let $u \in W^{1,1}(\Omega; \mathbb{R}^d)$, with $\nabla u \in BV(\Omega; \{A, B\})$. Then

$$\Gamma - \lim_{\varepsilon \to 0^+} I_\varepsilon(u; \Omega) = K_{\text{per}} \operatorname{Per}_\Omega(E),$$

where $\nabla u(x) = (1 - \chi_E(x)) A + \chi_E(x) B$ *for* $\mathcal{L}^N$ *a.e.* $x \in \Omega$, *and*

$$K_{\text{per}} := \inf \left\{ \int_Q L\, W(\nabla v) + \frac{1}{L} |\nabla^2 v|^2 \, dx : L > 0, v \in W^{2,\infty}(Q; \mathbb{R}^d), \right.$$

$$\left. \nabla v = \pm a \otimes e_N \text{ nearby } x_N = \pm \frac{1}{2}, \ v \text{ periodic of period one in } x' \right\}.$$

Note that the hypotheses of Theorems 2.1 and 2.2 are satisfied by the proto-type bulk energy density

$$W(\xi) := \min \left\{ |\xi - A|^2, |\xi - B|^2 \right\}.$$

As asserted above, under hypothesis (H_3) one has $K_{\text{per}} = K$, hence one-dimensional interface profiles are energetically preferred. Building upon the example by Jin and Kohn [90], in [44] we show that without hypothesis (H_3) we may have

$$K_{\text{per}} < K$$

thereby proving that optimal interface profiles are, at least in some cases, not one-dimensional. This happens because generating finite in-plane gradients (i.e. having a dependence on the coordinates parallel to the interface) reduces the energy in the regions far away from the two potential wells. The zero-curl constraint leads then to an oscillatory pattern. In elasticity, this multidimensional behavior has been predicted in [36]. Similar mechanisms are at play in the theory of micromagnetism, where indeed various non-one-dimensional wall structures are known, such as cross-tie domain walls and charged zigzag walls in ferromagnetic thin films (see e.g. [87] and references therein). It would be interesting to know if K_{per} is smaller or equal to K for realistic ferroelastic potentials obtained from the Landau theory of phase transitions.

3. Lower semicontinuity for higher order variational problems in $\mathbf{W}^{k,p}$, $\mathbf{p} > 1$

Morrey's notion of *quasiconvexity* was extended by Meyers [97] to the realm of higher-order variational problems. We recall that $f : \Omega \to [0, +\infty)$ is said to be *quasiconvex* if (see [46], [100])

$$f(\xi) \leq \int_Q f(\xi + \nabla \varphi(x)) \, dx \tag{8}$$

for all $\varphi \in C_c^\infty(Q; \mathbb{R}^d)$, and a function $F : E_k^d \to \mathbb{R}$ is said to be *k-quasiconvex* if

$$F(\xi) \leq \int_Q F(\xi + \nabla^k w(y)) \, dy \tag{9}$$

for all $\xi \in E_k^d$ and all $w \in C_c^\infty(Q; \mathbb{R}^d)$.

To fix notation, here and in what follows, Ω is an open, bounded domain in $\mathbb{R}^N$, $Q := (-1/2, 1/2)^N$, $C_c^\infty(\mathbb{R}^N; \mathbb{R}^d)$ is the space of infinitely differentiable $\mathbb{R}^d$-valued functions in Ω with compact support, and $C_{\mathrm{per}}^\infty(\mathbb{R}^N; \mathbb{R}^d)$ stands for the space of Q-periodic functions in $C^\infty(\mathbb{R}^N; \mathbb{R}^d)$. Recall that f is said to be Q-*periodic* if $f(x + k\mathbf{e}_i) = f(x)$ for all x, all $k \in \mathbb{Z}^k$, and for all $i = 1, \ldots, N$, where $\{\mathbf{e}_1, \ldots, \mathbf{e}_N\}$ is the standard orthonormal basis of $\mathbb{R}^N$. For each $j \in \mathbb{N}$ the symbol $\nabla^j u$ stands for the vector-valued function whose components are all the jth order derivatives of u. If u is C^∞ then for $j \geq 2$ we have that $\nabla^j u(x) \in E_j^d$, where E_j^d denotes the space of symmetric j-linear maps from $\mathbb{R}^N$ into $\mathbb{R}^d$. We set $E_0^d := \mathbb{R}^d$, $F_1^d := \mathbb{R}^{d \times N}$ and

$$E_{[j-1]}^d := E_0^d \times \cdots \times E_{j-1}^d, \quad E_{[0]}^d := E_0^d.$$

For any integer $k \geq 2$ we define

$$BV^k(\Omega; \mathbb{R}^d) := \left\{ u \in W^{k-1,1}(\Omega; \mathbb{R}^d) : \nabla^{k-1} u \in BV(\Omega; E_{k-1}^d) \right\},$$

where here $\nabla^j u$ is the Radon-Nikodym derivative of the distributional derivative $D^j u$ of $\nabla^{j-1} u$, with respect to the N-dimensional Lebesgue measure $\mathcal{L}^N$ (see [12]).

Meyers [97] proved that k-*quasiconvexity* is a necessary and sufficient condition for (sequential) lower semicontinuity of (1) with respect to weak convergence (resp. weak* convergence if $p = \infty$) in $W^{k,p}(\Omega; \mathbb{R}^d)$ under appropriate growth and continuity conditions on the integrand f. Meyers' argument uses results of Agmon, Douglis and Nirenberg [2] concerning Poisson kernels for elliptic equations, and later Fusco [80] gave a simpler proof using De Giorgi's Slicing Lemma. He also extended the result to Carathéodory integrands when $p = 1$, while the case $p > 1$ has been recently established by Guidorzi and Poggiolini [84], who relied heavily on a p-Lipschitz assumption, i.e.,

$$|f(x, \mathbf{v}, \xi) - f(x, \mathbf{v}, \xi_1)| \leq C(1 + |\xi|^{p-1} + |\xi_1|^{p-1})|\xi - \xi_1|.$$

As it turns out, k-quasiconvex integrands with p-growth are p-Lipschitz. This assertion was established by Marcellini [94] for $k = 1$, the case $k = 2$ was proven in [84], and recently Santos and Zappale [107] extended it to arbitrary k.

The first integral representations for the relaxed energies when the integrand depends on the full set of variables, that is $f = f(x, u, \ldots, \nabla^k u)$, were obtained by Braides, Fonseca and Leoni in [32], where this question may be seen as a corollary of very broad results casted for variational problems under PDE constraints (here curl= 0), the $\mathcal{A}$-*quasiconvexity theory*. In [74] Fonseca and Müller proved that $\mathcal{A}$-quasiconvexity is a necessary and sufficient condition for (sequential) lower semicontinuity of a functional

$$(u, v) \mapsto \int_\Omega f(x, u(x), v(x)) \, dx,$$

whenever $f : \Omega \times \mathbb{R}^d \times \mathbb{R}^m \to [0, \infty)$ is a Carathéodory integrand satisfying

$$0 \leq f(x, u, v) \leq a(x, u)(1 + |v|^q),$$

for a.e. $x \in \Omega$ and all $(u, v) \in \mathbb{R}^d \times \mathbb{R}^m$, where $1 \leq q < \infty$, $a \in L^\infty_{\text{loc}}(\Omega \times \mathbb{R}; [0, \infty))$, $u_n \to u$ in measure, $v_n \rightharpoonup v$ in $L^q(\Omega; \mathbb{R}^m)$ and $\mathcal{A}v_n \to 0$ in $W^{-1,q}(\Omega; \mathbb{R}^l)$ (see also [47]). In the sequel $\mathcal{A} : L^q(\Omega; \mathbb{R}^m) \to W^{-1,q}(\Omega; \mathbb{R}^l)$, $\mathcal{A}v := \sum_{i=1}^N A^{(i)} \frac{\partial v}{\partial x_i}$ is a constant-rank, first order linear partial differential operator, with $A^{(i)} : \mathbb{R}^m \to \mathbb{R}^l$ linear transformations, $i = 1, \ldots, N$. We recall that $\mathcal{A}$ satisfies the *constant-rank* property if there exists $r \in \mathbb{N}$ such that

$$\text{rank} \, \mathbb{A}w = r \quad \text{for all } w \in S^{N-1}, \tag{10}$$

where

$$\mathbb{A}w := \sum_{i=1}^N w_i A^{(i)}, \quad \text{for for} \quad w \in \mathbb{R}^N.$$

A function $f : \mathbb{R}^m \to \mathbb{R}$ is said to be $\mathcal{A}$-*quasiconvex* if

$$f(v) \leq \int_Q f(v + w(y)) \, dy$$

for all $v \in \mathbb{R}^m$ and all $w \in C^\infty_{\text{per}}(\mathbb{R}^N; \mathbb{R}^m)$ such that $\mathbb{A}w = 0$ and $\int_Q w(y) \, dy = 0$.

The relevance of this general framework, as emphasized by Tartar (see [113], [114], [115], [116], [117], [118]; see also [46], [47], [102]) lies on the fact that in continuum mechanics and electromagnetism PDEs other than $\text{curl} \, v = 0$ arise naturally, and this calls for a relaxation theory which encompasses PDE constraints of the type $\mathcal{A}v = 0$. Important examples include the cases where $\mathcal{A}v = 0$ if and only if $\text{curl} \, v = 0$, as well as k-th order gradients, where, by replacing the target space $\mathbb{R}^m$ by the finite dimensional vector space E_k^d, it is possible to find a first order linear partial differential operator $\mathcal{A}$ such that $v \in L^q(\Omega; E_k^d)$ and $\mathcal{A}v = 0$ if and only if there exists $\varphi \in W^{k,q}(\Omega; \mathbb{R}^d)$ such that $v = \nabla^k \varphi$ (see Corollary 3.2). Here A-quasiconvexity reduces to k-quasiconvexity (see (9)) when the energy density is continuous.

Let $1 \leq p < \infty$ and $1 < q < \infty$, and consider the functional

$$F : L^p(\Omega; \mathbb{R}^d) \times L^q(\Omega; \mathbb{R}^m) \times \mathcal{O}(\Omega) \to [0, \infty)$$

defined by

$$F((u, v); D) := \int_D f(x, u(x), v(x)) \, dx,$$

where $\mathcal{O}(\Omega)$ is the collection of all open subsets of Ω, and the density f satisfies the following hypothesis:

(H) $f : \Omega \times \mathbb{R}^d \times \mathbb{R}^m \to [0, \infty)$ is a Carathéodory function, and

$$0 \leq f(x, u, v) \leq C(1 + |u|^p + |v|^q)$$

for a.e. $x \in \Omega$ and all $(u, v) \in \mathbb{R}^d \times \mathbb{R}^m$, and for some constant $C > 0$.

For $D \in \mathcal{O}(\Omega)$ and $(u, v) \in L^p(\Omega; \mathbb{R}^d) \times (L^q(\Omega; \mathbb{R}^m) \cap \operatorname{Ker} \mathcal{A})$ define

$$\mathcal{F}((u,v); D) := \inf \Big\{ \liminf_{n\to\infty} F((u_n, v_n); D) : (u_n, v_n) \in L^p(D; \mathbb{R}^d) \times L^q(D; \mathbb{R}^m),$$

$$u_n \to u \text{ in } L^p(D; \mathbb{R}^d), \quad v_n \rightharpoonup v \text{ in } L^q(D; \mathbb{R}^m),$$

$$\mathcal{A}v_n \to 0 \text{ in } W^{-1,q}(D; \mathbb{R}^l) \Big\}.$$

(11)

It can be shown that the condition $\mathcal{A}v_n \to 0$ imposed in (11) may be equivalently replaced by requiring that v_n satisfy the homogeneous PDE $\mathcal{A}v = 0$. Precisely,

$$\mathcal{F}((u,v); D) = \inf \Big\{ \liminf_{n\to\infty} F((u_n, v_n); D) : (u_n, v_n) \in L^p(D; \mathbb{R}^d) \times L^q(D; \mathbb{R}^m),$$

$$u_n \to u \text{ in } L^p(D; \mathbb{R}^d), \quad v_n \rightharpoonup v \text{ in } L^q(D; \mathbb{R}^m), \quad \mathcal{A}v_n = 0 \Big\}.$$

(12)

The following integral representation for the relaxed energy $\mathcal{F}$ was obtained in [32].

Theorem 3.1 ([32], Theorem 1.1). *Under condition* (H) *and the constant-rank hypothesis* (10), *for all* $D \in \mathcal{O}(\Omega)$, $u \in L^p(\Omega; \mathbb{R}^d)$, *and* $v \in L^q(\Omega; \mathbb{R}^m) \cap \operatorname{Ker} \mathcal{A}$, *we have*

$$\mathcal{F}((u,v); D) = \int_D \mathcal{Q}_{\mathcal{A}} f(x, u(x), v(x)) \, dx$$

where, for each fixed $(x, u) \in \Omega \times \mathbb{R}^d$, *the function* $\mathcal{Q}_{\mathcal{A}} f(x, u, \cdot)$ *is the* $\mathcal{A}$-*quasiconvexification of* $f(x, u, \cdot)$, *namely*

$$\mathcal{Q}_{\mathcal{A}} f(x, u, v) := \inf \Big\{ \int_Q f(x, u, v + w(y)) \, dy : w \in C^\infty_{\mathrm{per}}(\mathbb{R}^N; \mathbb{R}^m) \cap \operatorname{Ker} \mathcal{A},$$

$$\int_Q w(y) \, dy = 0 \Big\}$$

for all $v \in \mathbb{R}^m$.

The proof of this theorem relies heavily on the use of Young measures (see [17], [121]), together with the blow-up method introduced by Fonseca and Müller in [72], and the arguments developed in [74] (see also [15], [93]) .

Corollary 3.2 ([32], Theorem 1.3). *Let* $1 \le p \le \infty$, $s \in \mathbb{N}$, *and suppose that* $f : \Omega \times E^d_{[k-1]} \times E^d_k \to [0, \infty)$ *is a Carathéodory function satisfying*

$$0 \le f(x, \mathbf{u}, v) \le C(1 + |\mathbf{u}|^p + |v|^p), \qquad 1 \le p < \infty,$$

for a.e. $x \in \Omega$ *and all* $(\mathbf{u}, v) \in E^d_{[k-1]} \times E^d_k$, *where* $C > 0$, *and*

$$f \in L^\infty_{\mathrm{loc}}(\overline{\Omega} \times E^d_{[k-1]} \times E^d_k; [0, \infty)) \qquad \textit{if } p = \infty.$$

Then for every $u \in W^{k,p}(\Omega; \mathbb{R}^d)$ we have

$$\int_\Omega \mathcal{Q}^k f(x, u, \ldots, \nabla^k u)\, dx = \inf \left\{ \liminf_{n \to \infty} \int_\Omega f(x, u_n, \ldots, \nabla^k u_n)\, dx : \right.$$

$$\left. \{u_n\} \subset W^{k,p}(\Omega; \mathbb{R}^d), u_n \rightharpoonup u \text{ in } W^{k,p}(\Omega; \mathbb{R}^d) \quad (\overset{*}{\rightharpoonup} \text{ if } p = \infty) \right\},$$

where, for a.e. $x \in \Omega$ and all $(\mathbf{u}, v) \in E^d_{[k-1]} \times E^d_k$,

$$\mathcal{Q}^k f(x, \mathbf{u}, v) := \inf \left\{ \int_Q f(x, \mathbf{u}, v + \nabla^k w(y))\, dy : w \in C^\infty_{\text{per}}(\mathbb{R}^N; \mathbb{R}^d) \right\}.$$

4. Lower semicontinuity for second order variational problems in BH^p, p > 1

There are now several lower semicontinuity results in spaces involving jumps as well as higher order derivatives. In particular, in [70] we considered functionals defined on the space of *functions with bounded Hessian*

$$\begin{aligned} BH(\Omega; \mathbb{R}^d) &:= \{u \in W^{1,1}(\Omega; \mathbb{R}^d) : D^2 u \text{ is a finite Radon measure}\} \\ &= \{u \in L^1(\Omega; \mathbb{R}^d) : Du \in BV(\Omega; \mathbb{R}^{d \times N})\} \end{aligned}$$

where $D^2 u$ denotes the distributional Hessian of u. For $1 < p < +\infty$ we also define

$$BH^p(\Omega; \mathbb{R}^d) := \{u \in BH(\Omega; \mathbb{R}^d) : \nabla^2 u \in L^p(\Omega; E^d_2)\}.$$

For various properties of the space BH, we refer to Demengel [55], [56], Carriero, Leaci and Tomarelli [38] and Temam [119]. We recall that if $u \in BH(\Omega; \mathbb{R}^d))$ then $Du = \nabla u$ and $[\nabla u](x) = 0$ for $\mathcal{H}^{N-1}$-a.e. $x \in \Omega$. We obtained the following BH generalization of Ambrosio's Theorem 4.3 in [8] originally stated within the framework of SBV spaces.

Theorem 4.1 ([70], Theorem 1.2). *Let $\Omega \subset \mathbb{R}^N$ be an open bounded set and let*

$$f : \Omega \times \mathbb{R}^N \times \mathbb{R}^{d \times N} \times E^d_2 \to [0, +\infty)$$

be a normal integrand, 2-quasiconvex in Λ, and such that

$$|\Lambda|^p \leq f(x, u, \xi, \Lambda) \leq a(x, u, \xi)(1 + |\Lambda|^p) \tag{13}$$

for $\mathcal{L}^N$ a.e. $x \in \Omega$ and all $(u, \xi, \Lambda) \in \mathbb{R}^N \times \mathbb{R}^{d \times N} \times E^d_2$, where $a(x, u, \xi)$ is a non-negative constant, and $p > 1$. Then for every $u \in W^{1,1}(\Omega; \mathbb{R}^d)$ and any sequence $\{u_n\} \subset BH(\Omega; \mathbb{R}^d)$ converging to u in $W^{1,1}(\Omega; \mathbb{R}^d)$ and such that

$$|D^2_s u_n|(\Omega) \to 0 \tag{14}$$

we have

$$\int_\Omega f(x, u, \nabla u, \nabla^2 u)\, dx \leq \liminf_{n \to \infty} \int_\Omega f(x, u_n, \nabla u_n, \nabla^2 u_n)\, dx.$$

We remark that no smoothness nor integrability properties are required from the function $(x, u, \xi) \mapsto a(x, u, \xi)$ that appears in the upper bound (13). All that is needed is that $a(x, u, \xi)$ be defined, finite, and non-negative for $\mathcal{L}^N$ a.e. $x \in \Omega$ and all $(u, \xi, \Lambda) \in \mathbb{R}^N \times \mathbb{R}^{d \times N} \times E_2^d$. In this generality, Theorem 4.1 is new even in the Sobolev setting, considerably improving our Corollary 3.2 above. Indeed, we have

Corollary 4.2 ([70], Corollary 1.3). *Let $\Omega \subset \mathbb{R}^N$ be an open bounded set and let*

$$f : \Omega \times \mathbb{R}^N \times \mathbb{R}^{d \times N} \times E_2^d \to [0, +\infty)$$

be a normal integrand, 2-quasiconvex in Λ, and such that

$$0 \leq f(x, u, \xi, \Lambda) \leq a(x, u, \xi)(1 + |\Lambda|^p)$$

for $\mathcal{L}^N$ a.e. $x \in \Omega$ and all $(u, \xi, \Lambda) \in \mathbb{R}^N \times \mathbb{R}^{d \times N} \times E_2^d$, where $a(x, u, \xi)$ is a non-negative constant, and $p > 1$. Then for every $u \in W^{2,p}(\Omega; \mathbb{R}^d)$ and any sequence $\{u_n\} \subset W^{2,p}(\Omega; \mathbb{R}^d)$ weakly converging to u in $W^{2,p}(\Omega; \mathbb{R}^d)$ we have

$$\int_\Omega f(x, u, \nabla u, \nabla^2 u) \, dx \leq \liminf_{n \to \infty} \int_\Omega f(x, u_n, \nabla u_n, \nabla^2 u_n) \, dx.$$

The approach in Theorem 4.1 is quite different from the one of Theorem 3.1 and uses instead some approximation results of BH^p functions by $W^{2,\infty}$ functions obtained following Ambrosio [8] lead and Acerbi and Fusco [1] ideas, via maximal functions.

Indeed, the A-quasiconvexity method used in [32] strongly relies on the underlying PDEs which characterize the space $W^{2,p}(\Omega; \mathbb{R}^d)$. However, in a recent paper [69] extending a result of Alberti [3] we have shown that in the passage from the Sobolev spaces to the space BH the Hessian matrix $D^2 u$ remains symmetric but it may loose, in general, the PDE constraint curl= 0. More precisely, we have proved that

Theorem 4.3. *([69], Theorem 1.4) Let Ω be an open subset of $\mathbb{R}^N$ and let f be a function in $L^1(\Omega; E_2^d)$. Then there exists $u \in BH(\Omega)$ and a constant $C > 0$ depending only on N such that*

$$D^2 u = f \mathcal{L}^N + [\nabla u] \otimes \nu_{\nabla u} \mathcal{H}^{N-1} \lfloor S(\nabla u),$$

and

$$\int_\Omega |u| + |\nabla u| \, dx + \int_{S(\nabla u) \cap \Omega} |[\nabla u]| \, d\mathcal{H}^{N-1} \leq C \int_\Omega |f| \, dx.$$

5. Lower semicontinuity for higher order variational problems in $\mathbf{W}^{k,1}$

As mentioned before, classical truncation methods for $k = 1$ cannot be extended in a simple way to truncate higher order derivatives, and successful techniques often rely on p-equi-integrability, and thus cannot work in the linear growth case. Indeed, when $p = 1$ due to loss of reflexivity of the space $W^{k,1}(\Omega; \mathbb{R}^d)$ one can only conclude that an energy bounded sequence $\{u_n\} \subset W^{k,1}(\Omega; \mathbb{R}^d)$ with $\sup_n \|u_n\|_{W^{k,1}} < \infty$ admits a subsequence (not relabeled) such that

$$u_n \to u \quad \text{in } W^{k-1,1}(\Omega; \mathbb{R}^d),$$

where $u \in W^{k-1,1}(\Omega; \mathbb{R}^d)$ and $\nabla^{k-1} u$ is a vector-valued function of bounded variation. Consequently, we now seek to establish lower semicontinuity in the space $W^{k,1}(\Omega; \mathbb{R}^d)$ under this natural notion of convergence, and when $u \in BV^k(\Omega; \mathbb{R}^d)$ (see [61], [122]). When $k = 1$ the scalar case $d = 1$ has been extensively treated, while the vectorial case $d > 1$ was first studied by Fonseca and Müller in [72] where it was proven (sequential) lower semicontinuity in $W^{1,1}(\Omega; \mathbb{R}^d)$ of a functional

$$u \mapsto \int_\Omega f(x, u(x), \nabla u(x))\, dx,$$

with respect to strong convergence in $L^1(\Omega; \mathbb{R}^d)$ (see also [10], [65], [66], [73], and the references contained therein). The approach in [72] is based on blow-up and truncation methods.

The following theorem was proved in the case $k = 1$ by Ambrosio and Dal Maso [10], while Fonseca and Müller [72] treated general integrands of the form $f = f(x, u, \nabla u)$, but their argument requires coercivity (see also [81]). The case $k \geq 2$ is due to Amar and De Cicco [5] (see [68] for a proof for all $k \geq 1$).

Proposition 5.1 ([68], Proposition 2.1). *Let* $f : E_k^d \to [0, \infty)$ *be a function k-quasiconvex, such that*

$$0 \leq f(\xi) \leq C(1 + |\xi|),$$

for all $\xi \in E_k^d$. *Moreover, when* $k \geq 2$ *assume that*

$$f(\xi) \geq C_1 |\xi| \quad \text{for } |\xi| \text{ large.}$$

Let $\{u_n\}$ *be a sequence of functions in* $W^{k,1}(Q; \mathbb{R}^d)$ *converging to 0 in the space* $W^{k-1,1}(Q; \mathbb{R}^d)$. *Then*

$$f(0) \leq \liminf_{n \to \infty} \int_Q f(\nabla^k u_n)\, dx.$$

More generally we consider the case where f depends essentially only on x and on the highest order derivatives, that is $\nabla^k u(x)$. This situation is significantly simpler than the general case, since it does not require to truncate the initial sequence $\{u_n\} \subset W^{k,1}(\Omega; \mathbb{R}^d)$.

Theorem 5.2 ([68], Theorem 1.1). *Let* $f : \Omega \times E_{[k-1]}^d \times E_k^d \to [0, \infty)$ *be a Borel integrand. Suppose that for all* $(x_0, \mathbf{v}_0) \in \Omega \times E_{[k-1]}^d$ *and* $\varepsilon > 0$ *there exist* $\delta_0 > 0$ *and a modulus of continuity* ρ, *with* $\rho(s) \leq C_0(1 + s)$ *for* $s > 0$ *and for some* $C_0 > 0$, *such that*

$$f(x_0, \mathbf{v}_0, \xi) - f(x, \mathbf{v}, \xi) \leq \varepsilon(1 + f(x, \mathbf{v}, \xi)) + \rho(|\mathbf{v} - \mathbf{v}_0|) \tag{15}$$

for all $x \in \Omega$ with $|x - x_0| \le \delta_0$, and for all $(\mathbf{v}, \xi) \in E^d_{[k-1]} \times E^d_k$. Assume also that one of the following three conditions is satisfied:
(a) $f(x_0, \mathbf{v}_0, \cdot)$ is k-quasiconvex in E^d_k and

$$\frac{1}{C_1}|\xi| - C_1 \le f(x_0, \mathbf{v}_0, \xi) \le C_1(1 + |\xi|) \qquad \text{for all } \xi \in E^d_k,$$

where $C_1 > 0$;
(b) $f(x_0, \mathbf{v}_0, \cdot)$ is 1-quasiconvex in E^d_k and

$$0 \le f(x_0, \mathbf{v}_0, \xi) \le C_1(1 + |\xi|) \qquad \text{for all } \xi \in E^d_k,$$

where $C_1 > 0$;
(c) $f(x_0, \mathbf{v}_0, \cdot)$ is convex in E^d_k.
Let $u \in BV^k(\Omega; \mathbb{R}^d)$ and let $\{u_n\}$ be a sequence of functions in $W^{k,1}(\Omega; \mathbb{R}^d)$ converging to u in $W^{k-1,1}(\Omega; \mathbb{R}^d)$. Then

$$\int_\Omega f(x, u, \dots, \nabla^k u) \, dx \le \liminf_{n \to \infty} \int_\Omega f(x, u_n, \dots, \nabla^k u_n) \, dx.$$

Here $f(x_0, \mathbf{v}_0, \cdot)$ is said to be 1-quasiconvex if $f(x_0, \mathbf{v}_0, \cdot)$ is the trace on E^d_k of a 1-quasiconvex function $\bar{f}$ defined on $\mathbb{R}^{(d \times N^{k-1}) \times N}$.

An important class of integrands which satisfy (15) of Theorem 5.2 is given by

$$f = f(x, \xi) := h(x)g(\xi),$$

where $h(x)$ is a non-negative lower semicontinuous function and g is a non-negative function which satisfies either (a) or (b) or (c). The case where $h(x) \equiv 1$ and g satisfies condition (a) was proved by Amar and De Cicco [5]. Theorem 5.2 extends a result of Fonseca and Leoni (Theorem 1.7 in [65]) to higher order derivatives, where the statement is exactly that of Theorem 5.2 setting $k = 1$ and excluding part (a). Related results when $k = 1$ were obtained previously by Serrin [108] in the scalar case $d = 1$ and by Ambrosio and Dal Maso [10] in the vectorial case $d > 1$ (see also Fonseca and Müller [72], [73]). Even in the simple case where $f = f(\xi)$ it is not known if Theorem 5.2(a) still holds without the coercivity condition

$$f(\xi) \ge \frac{1}{C_1}|\xi| - C_1.$$

The main tool in the proof of Theorem 5.2, used also in an essential way in subsequent results, is the *blow-up method* introduced by Fonseca and Müller [72], [73], which reduces the domain Ω to a ball and the target function u to a polynomial.

When the integrand f depends on the full set of variables in an essential way, the situation becomes significantly more complicated since one needs to truncate gradients and higher order derivatives in order to localize lower order terms. The following theorem was proved for $k = 1$ by Fonseca and Leoni in [65], Theorem 1.8, and extended to the higher order case in [68].

Theorem 5.3 ([68], Theorem 1.2). *Let $f : \Omega \times E^d_{[k-1]} \times E^d_k \to [0, \infty)$ be a Borel integrand, with $f(x, \mathbf{v}, \cdot)$ 1-quasiconvex in E^d_k. Suppose that for all $(x_0, \mathbf{v}_0) \in \Omega \times E^d_{[k-1]}$ either $f(x_0, \mathbf{v}_0, \cdot) \equiv 0$, or for every $\varepsilon > 0$ there exist $C, \delta_0 > 0$ such that*

$$f(x_0, \mathbf{v}_0, \xi) - f(x, \mathbf{v}, \xi) \leq \varepsilon(1 + f(x, \mathbf{v}, \xi)), \tag{16}$$

$$C|\xi| - \frac{1}{C} \leq f(x_0, \mathbf{v}_0, \xi) \leq C(1 + |\xi|) \tag{17}$$

for all $(x, \mathbf{v}) \in \Omega \times E^d_{[k-1]}$ with $|x - x_0| + |\mathbf{v} - \mathbf{v}_0| \leq \delta_0$ and for all $\xi \in E^d_k$. Let $u \in BV^k(\Omega; \mathbb{R}^d)$, and let $\{u_n\}$ be a sequence of functions in $W^{k,1}(\Omega; \mathbb{R}^d)$ converging to u in $W^{k-1,1}(\Omega; \mathbb{R}^d)$. Then

$$\int_\Omega f(x, u, \dots, \nabla^k u) \, dx \leq \liminf_{n \to \infty} \int_\Omega f(x, u_n, \dots, \nabla^k u_n) \, dx.$$

A standing open problem is to decide whether Theorem 5.3 continues to hold under the weaker assumption that $f(x, \mathbf{v}, \cdot)$ is k-quasiconvex, which is the natural assumption in this context.

As in Theorem 5.2, conditions (16) and (17) can be considerably weakened if we assume that $f(x, \mathbf{v}, \cdot)$ is convex rather than 1-quasiconvex. Indeed we have the following result:

Theorem 5.4 ([68], Theorem 1.5). *Let $f : \Omega \times E^d_{[k-1]} \times E^d_k \to [0, \infty]$ be a lower semi-continuous function, with $f(x, \mathbf{v}, \cdot)$ convex in E^d_k. Suppose that for all $(x_0, \mathbf{v}_0) \in \Omega \times E^d_{[k-1]}$ either $f(x_0, \mathbf{v}_0, \cdot) \equiv 0$, or there exist $C_1, \delta_0 > 0$, and a continuous function $g : B(x_0, \delta_0) \times B(\mathbf{v}_0, \delta_0) \to E^d_k$ such that*

$$f(x, \mathbf{v}, g(x, \mathbf{v})) \in L^\infty \left(B(x_0, \delta_0) \times B(\mathbf{v}_0, \delta_0); \mathbb{R} \right), \tag{18}$$

$$f(x, \mathbf{v}, \xi) \geq C_1 |\xi| - \frac{1}{C_1}$$

for all $(x, \mathbf{v}) \in \Omega \times E^d_{[k-1]}$ with $|x - x_0| + |\mathbf{v} - \mathbf{v}_0| \leq \delta_0$ and for all $\xi \in E^d_k$. Let $u \in BV^k(\Omega; \mathbb{R}^d)$, and let $\{u_n\}$ be a sequence of functions in $W^{k,1}(\Omega; \mathbb{R}^d)$ converging to u in $W^{k-1,1}(\Omega; \mathbb{R}^d)$. Then

$$\int_\Omega f(x, u, \dots, \nabla^k u) \, dx \leq \liminf_{n \to \infty} \int_\Omega f(x, u_n, \dots, \nabla^k u_n) \, dx.$$

Theorem 5.4 was obtained by Fonseca and Leoni (see [66], Theorem 1.1) in the case $k = 1$. It is interesting to observe that without a condition of the type (18) Theorem 5.4 is false in general. This has been recently proved by Černý and Malý in [41].

The proofs of Theorems 5.2(b) and (c), 5.3 and 5.4 can be deduced easily from the corresponding ones in [65], [66], where $k = 1$. It suffices to write

$$\int_\Omega f(x, u(x), \dots, \nabla^k u(x)) \, dx =: \int_\Omega F(x, \mathbf{v}(x), \nabla \mathbf{v}(x)) \, dx$$

with $\mathbf{v} := (u, \ldots, \nabla^{k-1}u)$, and then to perturb the new integrand F in order to recover the full coercivity conditions necessary to apply the results in [65], [66]. This approach cannot be used for k-polyconvex integrands and a new proof is needed to treat this case (see [52]). Thus Theorem 5.2(a) and Theorem 5.5 below are the only truly genuine higher order results, in that they cannot be reduced in a trivial way to a first order problem.

For each $\xi \in E_k^d$ let $\mathcal{M}(\xi) \in \mathbb{R}^\tau$ be the vector whose components are all the minors of ξ.

Theorem 5.5 ([65], Theorem 1.6). *Let $h : \Omega \times E_{[k-1]}^d \times \mathbb{R}^\tau \to [0, \infty]$ be a lower semicontinuous function, with $h(x, \mathbf{v}, \cdot)$ convex in $\mathbb{R}^\tau$. Suppose that for all points $(x_0, \mathbf{v}_0) \in \Omega \times E_{[k-1]}^d$ either $h(x_0, \mathbf{v}_0, \cdot) \equiv 0$, or there exist C, $\delta_0 > 0$, and a continuous function $g : B(x_0, \delta_0) \times B(\mathbf{v}_0, \delta_0) \to \mathbb{R}^\tau$ such that*

$$h(x, \mathbf{v}, g(x, \mathbf{v})) \in L^\infty \left(B(x_0, \delta_0) \times B(\mathbf{v}_0, \delta_0); \mathbb{R} \right),$$

$$h(x, \mathbf{v}, v) \geq C|v| - \frac{1}{C}$$

for all $(x, \mathbf{v}) \in \Omega \times E_{[k-1]}^d$ with $|x - x_0| + |\mathbf{v} - \mathbf{v}_0| \leq \delta_0$ and for all $v \in \mathbb{R}^\tau$. Let $u \in BV^k(\Omega; \mathbb{R}^d)$, and let $\{u_n\}$ be a sequence of functions in $W^{k,p}(\Omega; \mathbb{R}^d)$ which converges to u in $W^{k-1,1}(\Omega; \mathbb{R}^d)$, where p is the minimum between N and the dimension of the vectorial space E_{k-1}^d. Then

$$\int_\Omega h\left(x, u, \ldots, \nabla^{k-1}u, \mathcal{M}(\nabla^k u)\right) dx$$

$$\leq \liminf_{n \to \infty} \int_\Omega h(x, u_n, \ldots, \nabla^{k-1}u_n, \mathcal{M}(\nabla^k u_n)) \, dx.$$

Theorem 5.5 is closely related to a result of Ball, Currie and Olver [18], where it was assumed that

$$h(x, \mathbf{v}, v) \geq \gamma\left(|v|\right) - \frac{1}{C},$$

where

$$\frac{\gamma(s)}{s} \to \infty \text{ as } s \to \infty.$$

Also, as stated above and with $k = 1$, Theorem 5.5 was proved by Fonseca and Leoni in [66], Theorem 1.4.

In the scalar case $d = 1$, that is when u is an $\mathbb{R}$-valued function, and for first order gradients, i.e. $k = 1$, condition (17) can be eliminated, see Theorem 1.1 in [65]. In particular in [65] Fonseca and Leoni have shown the following result

Proposition 5.6 ([65], Corollary 1.2). *Let $g : \mathbb{R}^N \to [0, \infty)$ be a convex function, and let $h : \Omega \times \mathbb{R} \to [0, \infty)$ be a lower semicontinuous function. If $u \in BV(\Omega; \mathbb{R})$ and $\{u_n\} \subset W^{1,1}(\Omega; \mathbb{R})$ converges to u in $L^1(\Omega; \mathbb{R})$, then*

$$\int_\Omega h(x, u)g(\nabla u) \, dx \leq \liminf_{n \to \infty} \int_\Omega h(x, u_n)g(\nabla u_n) \, dx.$$

It is interesting to observe that an analog of this result is false when $k \geq 2$.

Theorem 5.7 ([68], Theorem 1.4). *Let $\Omega := (0,1)^N$, $N \geq 3$, and let h be a smooth cut-off function on $\mathbb{R}$ with $0 \leq h \leq 1$, $h(u) = 1$ for $u \leq \frac{1}{2}$, $h(u) = 0$ for $u \geq 1$. There exists a sequence of functions $\{u_n\}$ in $W^{2,1}(\Omega;\mathbb{R})$ converging to zero in $W^{1,1}(\Omega;\mathbb{R})$ such that $\{\|\Delta u_n\|_{L^1(\Omega;\mathbb{R})}\}$ is uniformly bounded and*

$$\limsup_{n\to\infty} \int_\Omega h(u_n)(1 - \Delta u_n)^+ \, dx < \int_\Omega h(0) \, dx.$$

Once again, we are confronted here with new challenges that are present on variational problems involving higher order derivatives.

Acknowledgement. This material is based upon work supported by the National Science Foundation under Grants No. DMS–9731957, DMS–0103798, by the Italian MURST, and through the Center for Nonlinear Analysis under NSF Grant No. DMS–9803791.

References

[1] E. Acerbi and N. Fusco, *Semicontinuity problems in the calculus of variations*, Arch. Rat. Mech. Anal., **86** (1984), 125–145.

[2] S. Agmon, A. Douglis and L. Nirenberg, *Estimates near the boundary for solutions of elliptic partial differential equations satisfying general boundary conditions. I*, Comm. Pure Applied Math., **12** (1959), 623–727.

[3] G. Alberti, *A Lusin type theorem for gradients*, Funct. Anal., **100** (1991), 110–118.

[4] G. Alberti, *Variational models for phase transitions, an approach via Γ-convergence*, Quaderni del Dipartimento di Matematica "U. Dini", Università degli Studi di Pisa, 1998.

[5] M. Amar and V. De Cicco, *Relaxation of quasi-convex integrals of arbitrary order*, Proc. Roy. Soc. Edin., **124** (1994), 927–946.

[6] L. Ambrosio, *New lower semicontinuity results for integral functionals*, Rend. Accad. Naz. Sci. XL, **11** (1987) 1–42.

[7] L. Ambrosio, *Metric space valued functions of bounded variation*, Ann. Scuola Norm. Sup. Pisa Cl. Sci., **17** (1990), 439–478.

[8] L. Ambrosio, *On the lower semicontinuity of quasiconvex integrals in* SBV$(\Omega, \mathbb{R}^k)$, Nonlinear Anal., **23** (1994), 405–425.

[9] L. Ambrosio and A. Braides, *Energies in SBV and variational models in fracture mechanics. Homogenization and applications to material sciences* (Nice, 1995), 1–22, GAKUTO Internat. Ser. Math. Sci. Appl., **9**, Tokyo, 1995.

[10] L. Ambrosio and G. Dal Maso, *On the relaxation in* $BV(\Omega;\mathbb{R}^m)$ *of quasi-convex integrals*, J. Funct. Anal., **109** (1992) 76–97.

[11] L. Ambrosio, C. DeLellis, and C. Mantegazza, *Line energies for gradient vector fields in the plane*, Calc. Var. Partial Differential Equations, **9** (1999), 327–355.

[12] L. AMBROSIO, N. FUSCO AND D. PALLARA, *Functions of Bounded Variation and Free Discontinuity Problems*, Mathematical Monographs, Oxford University Press, 2000.

[13] P. Aviles and Y. Giga, *A mathematical problem related to the physical theory of liquid crystal configurations*, Proc. Centre Math. Anal. Austr. Nat. Univ., **12** (1987), 1–16.

[14] P. Aviles and Y. Giga, *On lower semicontinuity of a defect energy obtained by a singular limit of the Ginzburg-Landau type energy for gradient fields*, Proc. Roy. Soc. Edin. Sect. A, **129** (1999), 1–17.

[15] E. J. Balder, *A general approach to lower semicontinuity and lower closure in optimal control theory*, SIAM J. Control Opt., **22** (1984), 570–598.

[16] S. Baldo, *Minimal interface criterion for phase transitions in mixtures of Cahn-Hilliard fluids*, Ann. Inst. H. Poincaré, Anal. Non Linéaire, **7** (1990), 67–90.

[17] J. M. Ball, *A version of the fundamental theorem for Young measures*, in PDE's and Continuum Models of Phase Transitions, M. Rascle, D. Serre, and M. Slemrod, eds., Lecture Notes in Physics, Vol. 344, Springer-Verlag, Berlin, 1989, 207–215.

[18] J. Ball, J. Currie and P. Olver, *Null lagrangians, weak continuity, and variational problems of arbitrary order*, J. Funct. Anal., **41** (1981), 315–328.

[19] J. Ball and R. D. James, *Fine phase mixtures as minimizers of energy*, Arch. Rat. Mech. Anal., **100** (1987), 13–52.

[20] A. C. Barroso and I. Fonseca, *Anisotropic singular perturbations – the vectorial case*, Proc. Roy. Soc. Edin. Sect. A, **124** (1994), 527–571.

[21] G. R. Barsch and J. A. Krumhansl, *Twin boundaries in ferroelastic media without interface dislocations*, Phys. Rev. Lett., **53** (1984), 1069–1072.

[22] H. Berliocchi and J. M. Lasry, *Intégrands normales et mesures paramétrées en calcul des variations*, Bull. Soc. Math. France, **101** (1973), 129–184.

[23] T. Bhattacharya and F. Leonetti, *A new Poincaré inequality and its application to the regularity of minimizers of integral functionals with non-standard growth*, Nonlinear Anal., **17** (1991), 833–839.

[24] E. Bombieri and E. Giusti, *A Harnack's type inequality for elliptic equations on minimal surfaces*, Inv. Math., **15** (1972), 24–46.

[25] G. Bouchitté, *Singular perturbations of variational problems arising from a two-phase transition model*, Appl. Math. Optim., **21** (1990), 289–314.

[26] B. Bourdin, G. A. Francfort and J.-J. Marigo, *Numerical experiments in revisited brittle fracture*, J. Mech. Phys. Solids, **48** (2000), 797–826.

[27] A. Braides, *A homogenization theorem for weakly almost periodic functionals*, Rend. Accad. Naz. Sci. XL, **104** (1986), 261–281.

[28] A. Braides, *Relaxation of functionals with constraints on the divergence*, Ann. Univ. Ferrara, Nuova Ser., Sez. VII 33 (1987), 157–177.

[29] A. Braides and A. Defranceschi, *Homogenization of Multiple Integrals*. Clarendon Press, Oxford, 1998.

[30] A. Braides and I. Fonseca, *Brittle thin films*, Appl. Math. Optim., **44** (2001), 299–323.

[31] A. Braides, I. Fonseca and G. Francfort, *3D-2D asymptotic analysis for inhomogeneous thin films*, Indiana U. Math J., **49** (2000), 1367–1404.

[32] A. Braides, I. Fonseca and G. Leoni, *A-quasiconvexity: relaxation and homogenization*, ESAIM Control Optim. Calc. Var., **5** (2000), 539–577.

[33] W. F. Brown, *Micromagnetics.* John Wiley and Sons, 1963.

[34] G. Buttazzo, *Semicontinuity, relaxation and integral representation problems in the Calculus of Variations.* Pitman Res. Notes in Math. 207, Longman, Harlow, 1989.

[35] J. W Cahn and J. E. Hilliard, *Free energy of a non-uniform system. I. Interfacial free energy*, J. Chem. Phys., **28** (1958), 258–267.

[36] W. Cao, G. R. Barsch, and J. A. Krumhansl, *Quasi-one-dimensional solutions for domain walls and their constraints in improper ferroelastics*, Phys. Rev. B, **42** (1990), 6396–6401.

[37] J. Carr, M. Gurtin, M. Slemrod, *Structured phase transitions on a finite interval*, Arch. Rational Mech. Anal., **86** (1984), 317–351.

[38] M. Carriero, A. Leaci and F. Tomarelli, *Special bounded hessian and elastic-plastic plate*, Rend. Accad. Naz. Sci. XL Mem. Mat., (5) **16** (1992), 223–258.

[39] M. Carriero, A. Leaci and F. Tomarelli, *Strong minimizers of Blake & Zisserman functional*, Ann. Sc. Norm. Sup. Pisa Cl. Sci., (4) **25** (1997), 257–285.

[40] M. Carriero, A. Leaci and F. Tomarelli, *A second order model in image segmentation: Blake & Zisserman functional*, Progr. Nonlinear Differential Equations Appl., 25, Birkhäuser **25** (1996), 57–72.

[41] R. Černý and J. Malý, *Counterexample to lower semicontinuity in Calculus of Variations.* To appear in Math. Z.

[42] R. Choksi, R. V. Kohn and F. Otto, *Domain branching in uniaxial ferromagnets: a scaling law for the minimum energy*, Comm. Math. Phys., **201** (1999), 61–79.

[43] B. Coleman, M. Marcus and V. Mizel, *On the thermodynamics of periodic phases*, Arch. Rat. Mech. Anal., **117** (1992), 321–347.

[44] S. Conti, I. Fonseca and G. Leoni, *A Γ-convergence result for the two-gradient theory of phase transitions*, submitted.

[45] S. H. Curnoe and A. E. Jacobs, *Twin wall of proper cubic-tetragonal ferroelastics*, Phys. Rev. B, **62** (2000), R11925–R11928.

[46] B. Dacorogna, *Direct methods in the calculus of variations*, Springer-Verlag, New York, 1989.

[47] B. Dacorogna, *Weak Continuity and Weak Lower Semicontinuity for Nonlinear Functionals*, Springer Lecture Notes in Mathematics, Vol 922, Springer-Verlag, Berlin, 1982.

[48] B. Dacorogna and I. Fonseca, *Minima Absolus pour des Energies Ferromagnétiques*, Comptes R. Ac. Sc. Paris, **331** (2000), 497–500.

[49] B. Dacorogna and I. Fonseca, *A-B Quasiconvexity and Implicit Partial Differential Equations*, to appear in Calc. Var.

[50] G. Dal Maso, *An Introduction to Γ-Convergence.* Birkhäuser, Boston, 1993.

[51] G. Dal Maso, A. Defranceschi and E. Vitali, *Private communication.*

[52] G. Dal Maso and C. Sbordone, *Weak lower semicontinuity of polyconvex integrals: a borderline case*, Math. Z. **218** (1995), 603–609.

[53] E. De Giorgi, *Sulla convergenza di alcune successioni di integrali del tipo dell'area*, Rend. Mat. (IV), **8** (1975), 277–294.

[54] De Giorgi and L. Ambrosio, *Un nuovo tipo di funzionale del calcolo delle variazioni*, Atti Accad. Naz. Lincei Rend. Cl. Sci. Fis. Mat. Natur., (8) **82** (1988), 199–210.

[55] F. Demengel, *Fonctions à hessien borné*, Ann. Inst. Fourier (Grenoble), **34** (1984), 155–190.

[56] F. Demengel, *Compactness theorems for spaces of functions with bounded derivatives and applications to limit analysis problems in plasticity*, Arch. Rational Mech. Anal., **105** (1989), 123–161.

[57] A. DeSimone, *Energy minimizers for large ferromagnetic bodies*, Arch. Rat. Mech. Anal., **125** (1993), 99–143.

[58] A. DeSimone, R.V. Kohn, S. Müller, F. Otto, *Magnetic microstructures – a paradigm of multiscale problems*. MPI Preprint .70/1999.

[59] A. DeSimone, R. V. Kohn, S. Müller, and F. Otto, *A compactness result in the gradient theory of phase transitions*, Proc. Roy. Soc. Edinburgh Sect. A, **131** (2001), 833–844.

[60] A. DeSimone, R. V. Kohn, S. Müller, F. Otto, R. Schäfer, *Two-Dimensional Modeling of Soft Ferromagnetic Films*. MPI Preprint 30/2000.

[61] L. C. Evans and R. F. Gariepy, *Lecture Notes on Measure Theory and Fine Properties of Functions*. Studies in Advanced Math., CRC Press, 1992.

[62] I. Fonseca, *The lower quasiconvex envelope of the stored energy function for an elastic crystal*, J. Math. Pures et Appl., **67** (1988), 175–195.

[63] I. Fonseca, *Phase transitions of elastic solid materials*, Arch. Rat. Mech. Anal., **107** (1989), 195–223.

[64] I. Fonseca and G. A. Francfort, *Relaxation in BV versus quasiconvexification in $W^{1,p}$; a model for the interaction between fracture and damage*, Calc. Var. PDE, **3** (1995), 407–446.

[65] I. Fonseca and G. Leoni, *On lower semicontinuity and relaxation*, Proc. Royal Soc. Edin. Sect. A, **131** (2001), 519–565.

[66] I. Fonseca and G. Leoni, *Some remarks on lower semicontinuity*, Indiana Univ. Math. J., **49** (2000), 617–635.

[67] I. Fonseca and G. Leoni, *Relaxation results in micromagnetics*, Ricerche di Matematica, **XLIX** (2000), 269–304.

[68] I. Fonseca, G. Leoni, J. Malý, and R. Paroni, *A Note on Meyer's Theorem in $W^{k,1}$*, to appear in Trans. A. M. S.

[69] I. Fonseca, G. Leoni and R. Paroni, *On Hessian Matrices in the Space BH*, submitted.

[70] I. Fonseca, G. Leoni and R. Paroni, *On lower semicontinuity in BH^p and 2-quasiconvexification*, submitted.

[71] I. Fonseca and C. Mantegazza, *Second order singular perturbation models for phase transitions*, SIAM J. Math. Anal., **31** (2000), 1121–1143.

[72] I. Fonseca and S. Müller, *Quasi-convex integrands and lower semicontinuity in L^1*, SIAM J. Math. Anal., **23** (1992), 1081–1098.

[73] I. Fonseca and S. Müller, *Relaxation of quasiconvex functionals in* BV$(\Omega, \mathbb{R}^p)$ *for integrands* $f(x, u, \nabla u)$, Arch. Rat. Mech. Anal., **123** (1993), 1–49.

[74] Fonseca I. and S. Müller, *A-quasiconvexity, lower semicontinuity and Young measures*, SIAM J. Math. Anal., **30** (1999) 1355–1390.

[75] I. Fonseca and L. Tartar, *The gradient theory of phase transitions for systems with two potential wells*, Proc. Roy. Soc. Edin. Sect. A, **111** (1989), 89–102.

[76] G. Francfort and J.-J. Marigo, *Stable damage evolution in a brittle continuous medium*, European J. Mech. A Solids, **12** (1993), 149–189.

[77] G. Francfort and J.-J. Marigo, *Revisiting brittle fracture as an energy minimization problem*, J. Mech. Phys. Solids, **46** (1998), 1319–1342.

[78] G. Francfort and J.-J. Marigo, *Cracks in fracture mechanics: a time indexed family of energy minimizers. Variations of domain and free-boundary problems in solid mechanics*, Solid Mech. Appl., **66**, Kluwer Acad. Publ., Dordrecht, 1999.

[79] G. Francfort and J.-J. Marigo, *Une approche variationnelle de la mécanique du défaut*. Actes du 3éme Congrès d'Analyse Numérique: CANum '98 (Arles, 1998), ESAIM Proc. **6**, 1999.

[80] N. Fusco, *Quasiconvessità e semicontinuità per integrali multipli di ordine superiore*, Ricerche Mat., **29** (1980), 307–323.

[81] N. Fusco and J. E. Hutchinson, *A direct proof for lower semicontinuity of polyconvex functionals*, Manus. Math., **85** (1995), 35–50.

[82] D. Gilbarg and N. S. Trudinger, *Elliptic partial differential equations of second order*. Springer-Verlag, Berlin, 2001.

[83] E. Gonzalez, U. Massari and I. Tamanini, *On the regularity of boundaries of sets minimizing perimeter with a volume constraint*, Indiana Univ. Math. J., **32** (1983), 25–37.

[84] M. Guidorzi and L. Poggiolini, *Lower semicontinuity for quasiconvex integrals of higher order*, NoDEA, **6** (1999), 227–246.

[85] M. Gurtin, *Two-phase deformations of elastic solids*, Arch. Rational Mech. Anal., **84** (1983), 1–29.

[86] M. E. Gurtin, *Some results and conjectures in the gradient theory of phase transitions*, IMA, Preprint 156, 1985.

[87] A. Hubert and R. Schäfer, *Magnetic domains: the analysis of magnetic microstructures*. Springer, Berlin, 1998.

[88] R. D. James and D. Kinderlehrer, *Frustation in ferromagnetic materials*, Continuum Mech. Thermodyn., **2** (1990), 215–239.

[89] R. D. James and S. Müller, *Internal variables and fine-scale oscillations in micromagnetics*, Continuum Mech. Thermodyn., **6** (1994), 291–336.

[90] W. Jin and R. V. Kohn, *Singular perturbation and the energy of folds*, J. Nonlinear Sci., **10** (2000), 355–390.

[91] R. V. Kohn and S. Müller, *Surface energy and microstructure in coherent phase transitions*, Comm. Pure Appl. Math., **47**, 405–435.

[92] R. V. Kohn and P. Sternberg, *Local minimisers and singular perturbations*, Proc. Roy. Soc. Edin. Sect. A, **111** (1989), 69–84.

[93] J. Kristensen, *Finite functionals and Young measures generated by gradients of Sobolev functions*, Mathematical Institute, Technical University of Denmark, Mat-Report No. 1994-34, 1994.

[94] P. Marcellini, *Approximation of quasiconvex functions and lower semicontinuity of multiple integrals quasiconvex integrals*, Manus. Math., **51** (1985), 1–28.

[95] P. Marcellini and C. Sbordone, *Semicontinuity problems in the Calculus of Variations*, Nonlinear Analysis, **4** (1980), 241–257.

[96] V.G. Maz'ja, *Sobolev spaces*. Springer-Verlag, Berlin, 1985.

[97] N. MEYERS, *Quasi-convexity and lower semi-continuity of multiple variational integrals of any order*, Trans. A. M. S. **119** (1965), 125–149.

[98] L. Modica, *The gradient theory of phase transitions and the minimal interface criterion*, Arch. Rat. Mech. Anal., **98** (1987), 123–142.

[99] L. Modica and S. Mortola, *Un esempio di Γ-convergenza*, Boll. Un. Mat. Ital. B, **14** (1977), 285–299.

[100] C. Morrey, *Quasi-convexity and the lower semicontinuity of multiple integrals*, Pacific J. Math., **2** (1952), 25–53.

[101] S. Müller *Variational models for microstructures and phase transitions*. Lecture Notes, MPI Leipzig, 1998.

[102] F. Murat, *Compacité par compensation: condition nécéssaire et suffisante de continuité faible sous une hypothèse de rang constant*, Ann. Sc. Norm. Sup. Pisa, **8** (1981), 68–102.

[103] M. Ortiz and G. Gioia, *The morphology and folding patterns of buckling-driven thin-film blisters*, J. Mech. Phys. Solids, **42** (1994), 531–559.

[104] N. Owen and P. Sternberg, *Nonconvex variational problems with anisotropic perturbations*, Nonlinear Anal., **16** (1991), 705–719.

[105] P. Pedregal, *Parametrized Measures and Variational Principles*. Birkhäuser, Boston, 1997.

[106] E. K. H. Salje, *Phase transitions in ferroelastic and co-elastic crystals*, Cambridge University Press, Cambridge, 1990.

[107] P. Santos and E. Zappale. *In preparation*.

[108] J. Serrin, *On the definition and properties of certain variational integrals*, Trans. A. M. S., **161** (1961), 139–167.

[109] E. A. Stein, *Singular Integrals and Differentiability Properties of Functions*. Princeton Univ. Press, Princeton, N. J., 1970.

[110] E. A. Stein and G. Weiss, *Introduction to Fourier Analysis on Euclidean Spaces*. Princeton Univ. Press, Princeton, N. J., 1971.

[111] P. Sternberg, *The effect of a singular perturbation on non-covex variational problems*, Arch. Rat. Mech. Anal., **101** (1988), 209–260.

[112] P. Sternberg, *Vector-valued local minimizers of non-convex variational problems*, Rocky Mountain J. Math., **21** (1991), 799–807.

[113] L. Tartar, *Compensated compactness and applications to partial differential equations*, in Nonlinear Analysis and Mechanics: Heriot-Watt Symposium, R. Knops, ed., vol. IV, Pitman Res. Notes Math. Vol 39, 1979, 136–212.

[114] L. Tartar, *The compensated compactness method applied to systems of conservation laws*, in Systems of Nonlinear Partial Differential Eq., J. M. Ball, ed., Riedel, 1983.

[115] L. Tartar, *Étude des oscillations dans les équations aux dérivées partielles non-linéaires*, Springer Lectures Notes in Physics, Springer-Verlag, Berlin, **195** (1984), 384–412.

[116] L. Tartar, *H-measures, a new approach for studying homogenisation, oscillations and concentration effects in partial differential equations*, Proc. Roy. Soc. Edin. Sect. A, **115A** (1990), 193–230.

[117] L. Tartar, *On mathematical tools for studying partial differential equations of continuum physics: H-measures and Young measures*, in Developments in Partial Differential Equations and Applications to Mathematical Physics, Buttazzo, Galdi, Zanghirati, eds., Plenum, New York, 1991.

[118] L. Tartar, *Some remarks on separately convex functions*, in Microstructure and Phase Transitions, D. Kinderlehrer, R. D. James, M. Luskin and J. L. Ericksen, eds., Vol. 54, IMA Vol. Math. Appl., Springer-Verlag, 1993, 191–204.

[119] T. Temam, *Problèmes mathématiques en plasticité*, Gauthier-Villars, Paris, 1983.

[120] A. Visintin, *On Landau-Lifschitz' equations for ferromagnetism*, Jap. J. Appl. Math., **2** (1985), 69–84.

[121] L. C. Young, *Lectures on Calculus of Variations and Optimal Control Theory*. W. B. Saunders, 1969.

[122] W.P. Ziemer, *Weakly differentiable functions. Sobolev spaces and functions of bounded variation*. Springer-Verlag, New York, 1989.

Irene Fonseca
Department of Mathematical Sciences
Carnegie Mellon University
Pittsburgh, PA 15213
USA
E-mail address: fonseca@andrew.cmu.edu

Giovanni Leoni
Dipartimento di Scienze e Tecnologie Avanzate
Università del Piemonte Orientale
I-15100 Alessandria
Italy
E-mail address: leoni@unipmn.it

Progress in Nonlinear Differential Equations
and Their Applications, Vol. 51, 141–153

Unstable Crystalline Wulff Shapes in 3D

Maurizio Paolini and Franco Pasquarelli

Abstract. We investigate the stability of the evolution by anysotropic and crystalline curvature starting from an initial surface equal to the Wulff shape. It is well known that the Wulff shape evolves selfsimilarly according to the law $V = -\kappa_\phi n_\phi$. Here the index ϕ refers to the underlying anisotropy described by the Wulff shape, so that κ_ϕ is the relative mean curvature and n_ϕ is the Cahn-Hoffmann conormal vector field. Such selfsimilar evolution is also known to be stable under small perturbations of the initial surface in the isotropic setting (the Wulff shape is a sphere) or in 2D if the underlying anisotropy is symmetric. We show that this evolution is unstable for some specific choices of the Wulff shape both rotationally symmetric and fully crystalline.

1. Introduction

Evolution of surfaces by their mean curvature has been deeply investigated in recent years, in its abstract mathematical aspects, see e.g. [12], [10], [14], [11], [7], [8], [19], and in applied and numerical fields [20], [15], [21], [13], [22]. One of the most famous results states that an evolving surface becomes more and more spherical before shrinking to a point and disappear. This is true in 2D for any embedded initial curve [17] and in higher dimensions if the initial hypersurface is convex [18].

The two-dimensional case has also been investigated, in this respect, in the anisotropic case by Gage [16] (for smooth and strictly convex anisotropies) and by A. Stancu [23] when the anisotropy is of crystalline type. If the metric describing the anisotropy is symmetric, they reach a conclusion similar to that of Grayson, exept for very few special crystalline anisotropies. Namely, they prove that for any smooth (resp. faceted in case of a crystalline anisotropy) initial convex curve, the corresponding evolution, when rescaled to constant enclosed area, approaches the so-called Wulff shape before disappearing.

Such result is believed to hold true in 2D also for a nonsymmetric anisotropy, although not yet proved, to our knowledge.

The situation is largely unknown in 3D. The techniques used by Gage and Stancu in their papers no longer apply in higher dimensions.

We show here that anisotropy in 3D can deeply change the behaviour of collapsing convex surfaces evolving by anisotropic mean curvature, and stability of the Wulff shape in the sense stated above cannot be expected, at least for a wide class of crystalline anisotropies; this is consistent with a result of J. Taylor [24]. We describe some examples of crystalline anisotropies for which even small convex perturbations of the Wulff shape lead to an evolution diverging from the Wulff shape. More precisely the surface tends to become more and more *flat* or more and more *elongated*, and in some cases it approaches a selfsimilarly evolving shape different from the Wulff shape itself.

2. Setting and notations

We briefly recall the notations and definitions of anisotropic motion by mean curvature. For the regural and strictly convex case they correspond to those in [9]. For a more complete and rigorous analysis of the crystalline case we refer to the papers [3] and [6].

Anisotropy itself is described by means of a norm $\phi : \mathbb{R}^3 \to \mathbb{R}^+$ which is intended to measure the length of vectors in the tangent space of any point $x \in \mathbb{R}^3$; the dual norm ϕ^o is then defined by

$$\phi^o(\xi^\star) := \max_{\xi : \phi(\xi) \leq 1} (\xi \cdot \xi^\star)$$

and is usually referred to as the surface energy density.

The unit balls $W_\phi := \{\xi : \phi(\xi) \leq 1\}$ and $F_\phi := \{\xi^\star : \phi^o(\xi^\star) \leq 1\}$ are the so-called Wulff shape and Frank diagram respectively. Crystalline anisotropy is characterized by a polyhedral Wulff shape; we shall also consider a mixed case where the Wulff shape is described by the rotation of a polygon around the vertical axis so that it is not regular nor polyhedral.

To simplify notation we shall drop from now on the superscript * for vectors in the cotangent space. Associated to ϕ^o we next define the duality operator

$$T^o(\xi) := \frac{1}{2} \nabla \left[\phi^o(\xi)\right]^2$$

where the gradient is to be understood in the sense of subdifferentials; it is a nonlinear maximal monotone and multivalued function $T^o : \mathbb{R}^3 \to \mathbb{R}^3$ which maps F_ϕ onto W_ϕ.

Let $\Sigma(t) = \partial A(t)$, $t \in [0, T)$ be an evolving oriented Lipschitz surface, possibly polyhedral in shape, and $\nu : \Sigma \to \mathbb{R}^3$ the corresponding outward euclidean normal vector field, which is well defined almost everywhere. We also introduce the normalized vector field $\nu_\phi = \frac{\nu}{\phi^o(\nu)}$. The so-called Cahn-Hoffmann vector field is then constructed so as to satisfy the relation

$$n_\phi \in T^o(\nu_\phi).$$

The actual definition of such vector field when T^o is multivalued is far from obvious and depends on the evolution itself. In the absence of a forcing term (which is the case here) this vector field solves a variational problem described in [4], [5]. We also refer to [2] for a more complete analysis.

Once we have the Cahn-Hoffmann field we are in a position to introduce the crystalline curvature using the surface divergence as follows

$$\kappa_\phi = \mathrm{div}_\Sigma\, n_\phi,$$

alternatively we can extend n_ϕ in a neighborhood of Σ with the restriction $\phi(n_\phi) = 1$ and then compute its divergence on Σ.

Finally, the evolution law reads as

$$V_\phi = -\kappa_\phi$$

where V_ϕ is the expanding velocity of Σ in the direction of ν_ϕ. It is related to the euclidean normal velocity which is given by $V_\nu = \phi^o(\nu)V_\phi$, so that $V_\nu = -\phi^o(\nu)\kappa_\phi$.

When $\Sigma(0) = \partial W_\phi$ we know the following facts [9]:

- The Cahn-Hoffmann vector field is given by $n_\phi(x) = x$, for any $x \in \partial W_\phi$;
- The crystalline curvature is constant $\kappa_\phi(x) = 2$, for any $x \in \partial W_\phi$;
- The evolution $\Sigma(t)$ originating from ∂W_ϕ is selfsimilar and given by $\Sigma(t) = \sqrt{1 - 4t}\,\partial W_\phi$.

3. First example: rotational shape

The Wulff shape W_ϕ of the anisotropy ϕ is depicted in Figure 1, where the vertical axis is actually a rotation axis. Quantities h and k have the meaning of a perturbation with respect to the cylindrical anisotropy having $B_1 \times\,]-1, 1[$ as Wulff shape, $B_1 \subseteq \mathbb{R}^2$ is the ball of radius 1 centered at the origin.

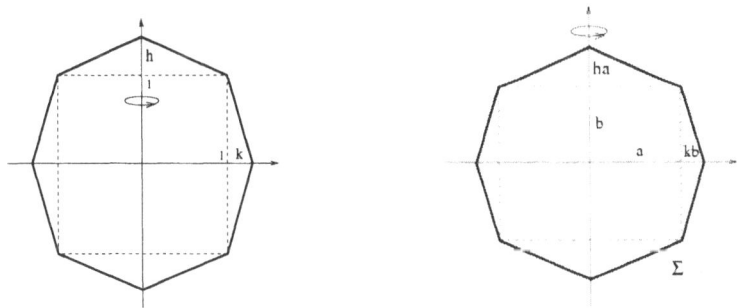

FIGURE 1. Section of the Wulff shape (left) and of the evolving surface $\Sigma(t)$ (right).

If $\xi = (x, y) \in \mathbb{R}^3$, $x \in \mathbb{R}^2$, we set $(r, y) = (|x|, y)$ and we shall compute the quantities $\phi(\xi)$ and $\phi^o(\xi)$ as functions of r and y, for simplicity we continue to denote those functions with ϕ and ϕ^o. We fix our attention to just some interesting directions. Recall that the dual norm ϕ^o can be computed by $\phi^o(\nu) = \max_{\{\phi(\xi)=1\}} \xi \cdot \nu$ and is of course homogeneous of degree 1. If ν is orthogonal to some side l of the

Wulff shape, than any choice of a point in l gives the maximum in the above definition of ϕ^o. It turns out that

$$\phi((1,0)) = \frac{1}{1+k}, \qquad \phi((1,1)) = 1, \qquad \phi((0,1)) = \frac{1}{1+h}, \tag{1}$$

$$\phi^o((1,k)) = (1,k) \cdot (1+k,0) = 1+k, \tag{2}$$

$$\phi^o((h,1)) = (h,1) \cdot (0,1+h) = 1+h. \tag{3}$$

We want to describe the evolution by crystalline anisotropy, as defined in [9], of an initial set $\Sigma(0)$ which is convex and admissible with respect to the given anisotropy. It turns out that such a set depends only on two quantities: $a(0)$ and $b(0)$ which are the radius and half the height of the inscribed cylinder. The surface is similar to the Wulff shape, with intersections with the coordinate axis at the points $(a + kb, 0)$ and $(0, b + hb)$, see Figure 1.

If the faces remain calibrable (see [3], [4], [5], [6]) during the evolution, then the evolving set $\Sigma(t)$ is also defined by two quantities $a(t)$ and $b(t)$, see Figure 1.

4. Computing the curvatures

For the top face (the lateral surface of a cone of radius a and height ha) the computation is easy, since this face is just a rescaling by a factor a and translation of the top face of W_ϕ, which has curvature 2 everywhere, so that

$$\kappa_{\text{TOP}} = \frac{2}{a}. \tag{4}$$

We now turn our attention to the lateral face. For a given time t set $A = (a + kb, 0)$ and $B = (a, b)$, they are two consecutive vertices of the polygonal defining $\Sigma(t)$ after rotation around the y-axis. On A and B the ϕ-normal vector is easily seen to be respectively $\tilde{n}_\phi(A) = (1 + k, 0)$ and $\tilde{n}_\phi(B) = (1, 1)$; we use the "tilde" because we are working in the (r, y) plane. On a generic point $P = \lambda B + (1 - \lambda)A$ of the segment AB, for $\lambda \in [0, 1]$, we only know that the ϕ-normal lies on the corresponding side of the Wulff shape (in the (r, y) plane), i.e. there exists a function $\mu : [0, 1] \to [0, 1]$, $\mu(0) = 0$, $\mu(1) = 1$ such that:

$$\tilde{n}_\phi(P) = \mu(\lambda)\tilde{n}_\phi(B) + (1 - \mu(\lambda))\tilde{n}_\phi(A).$$

The actual ϕ-normal vector field is obtained by rotation around the vertical axis and then should be extended outside $\Sigma(t)$ with the constraint of having unit ϕ-norm.

A quick computation shows that the divergence (in $\mathbb{R}^3$) of this vector field in P is given by

$$\kappa_\phi(P) = \text{div}_z\, n_\phi = \text{div}_{(r,y)}\, \tilde{n}_\phi + \frac{\tilde{n}_{\phi,1}}{P_1} \tag{5}$$

where the second term in the right-hand side of (5) comes from the rotation around the y-axis. The subscript 1 stands for the first component of $\tilde{n}_\phi$ and P respectively.

Since the component of $\tilde{n}_\phi$ normal to the line AB is constant, it follows that the divergence in $\mathbb{R}^2$ reduces to the tangential derivative of the tangential component of $\tilde{n}_\phi$ along AB:

$$\mathrm{div}_{(r,y)}\, \tilde{n}_\phi = \frac{\mathrm{d}\tilde{n}_\phi \cdot \tau}{\mathrm{d}\lambda}\frac{\mathrm{d}\lambda}{\mathrm{d}s} = \frac{1}{l}(1 + k^2)\mu'(\lambda)\frac{1}{bl} = \frac{\mu'(\lambda)}{b}$$

where $\tau = (B - A)/|AB|$, s parametrizes AB by arc length and l is the length of the side parallel to AB of the polygon generating the Wulff shape. This result is not surprising, since for the Wulff shape itself (where $\mu(\lambda) = \lambda$) it gives just 1, which is half of the expected value, the missing part coming from the rotation. The second term in (5) evaluates as follows

$$\frac{\tilde{n}_{\phi,1}}{P_1} = \frac{\mu(\lambda) + (1 - \mu(\lambda))(1 + k)}{\lambda a + (1 - \lambda)(a + kb)} = \frac{1 + k - \mu(\lambda)k}{a + kb - \lambda kb}.$$

As a verification, this expression evaluates to 1 if computed on the Wulff shape itself and by choosing the identity as μ.

The unknown function μ should now be chosen so as to give a constant κ_ϕ along AB. If we integrate κ_ϕ on the lateral surface we get (integating by parts)

$$\int_{\Sigma_{\mathrm{LAT}}} \kappa_\phi\, d\sigma = 2\pi \int_{AB} P_1(\lambda(s))\kappa_\phi(P)\, ds$$

$$= 2\pi bl \int_0^1 [P_1(\lambda)\frac{\mu'(\lambda)}{b} + 1 + k - \mu(\lambda)k]\, d\lambda$$

$$= 2\pi l \int_0^1 [\mu(\lambda)kb + b + kb - \mu(\lambda)kb]\, d\lambda + 2\pi l[\mu(\lambda)P_1(\lambda)]_0^1$$

$$= 2\pi l(b + kb + a).$$

Since $|\Sigma_{\mathrm{LAT}}| = \pi(2a + kb)lb$ we finally get the mean value of κ_ϕ and hence κ_ϕ:

$$\kappa_{\mathrm{LAT}} = 2\frac{a + b + kb}{b(2a + kb)} \tag{6}$$

which degenerates to $\frac{1}{a} + \frac{1}{b}$ when $k = 0$. It remains to determine μ by imposing that the crystalline curvature is exactly κ_{LAT} for any λ, we get (multiplying by b the two terms forming $\kappa_\phi(P)$)

$$\mu' + \frac{b + kb - \mu kb}{a + kb - \lambda kb} = 2\frac{a + b + kb}{2a + kb} \tag{7}$$

and we just need to check if $\mu(\lambda) \in [0, 1]$ for all $\lambda \in [0, 1]$, otherwise the lateral facet would not be calibrable and would break or bend during evolution. If $k = 0$ we immediately get $\mu' = 1$ and we obtain the calibrability of the lateral face.

Otherwise, if $k > 0$, by differentiating equation (7) with respect to λ, after some tedious computations we reach the conclusion that $\mu'' < 0$ whenever $\mu' = 0$, which means that no local minima is allowed for the function μ in $(0, 1)$. In other words, if $\mu'(1) \geq 0$ then we have calibrability of the lateral face; on the other hand, if $\mu'(1) < 0$ then the lateral face cannot be calibrable.

Now we can compute

$$\mu'(1) = \frac{2a^2 + 2kab - kb^2}{2a^2 + kab}$$

and the calibrability of the lateral face is equivalent to the non-negativity of the numerator:

$$2a^2 + 2kab - kb^2 \geq 0. \tag{8}$$

Unfortunately it is indeed possible that this quantity becomes negative during the evolution, when b becomes larger than $a + a\sqrt{1 + \frac{2}{k}}$. In such situation, the subsequent evolution should be computed in a different way. We shall not discuss this situation here.

5. Computing normal velocities

We now turn our attention to the ϕ-normal velocities of the top and lateral faces of $\Sigma(t)$, they will be functions of $a'(t)$ and $b'(t)$. We are interested in the velocity in the $\nu_\phi = \frac{\nu}{\phi^\circ(\nu)}$ direction, where ν here is any normal vector to Σ. If we fix our attention to some point $P(t) \in \Sigma(t)$, the required velocity is given by $P'(t) \cdot \nu_\phi$.

Now let $P(t) = (0, b + ha)$ (the top vertex of $\Sigma(t)$ in the (r, y) plane), and consider that the vector $\nu = (h, 1)$ is orthogonal to the top face of Σ, then, using (3)

$$V_{\text{TOP}} = \frac{(0, b' + ha') \cdot (h, 1)}{\phi^\circ((h, 1))} = \frac{b' + ha'}{1 + h} \tag{9}$$

and similarly, using (2), we get

$$V_{\text{LAT}} = \frac{(a' + kb', 0) \cdot (1, k)}{\phi^\circ((1, k))} = \frac{a' + kb'}{1 + k}. \tag{10}$$

We are finally in a position to enforce the relations $V_{\text{TOP}} = -\kappa_{\text{TOP}}$ and $V_{\text{LAT}} = -\kappa_{\text{LAT}}$. Using (9) and (10) we can eliminate first b' and then a' from such relations to get

$$\begin{aligned}(1 - hk)a' &= k(1 + h)\kappa_{\text{TOP}} - (1 + k)\kappa_{\text{LAT}} \\ (1 - hk)b' &= h(1 + k)\kappa_{\text{LAT}} - (1 + h)\kappa_{\text{TOP}}.\end{aligned} \tag{11}$$

6. Analysis of the system

Plugging (4) and (6) into (11) we get

$$\begin{cases} (1 - hk)a' &= \frac{2}{ab(2a+kb)}[k(1 + h)b(2a + kb) - (1 + k)a(a + b + kb)] \\ (1 - hk)b' &= \frac{2}{ab(2a+kb)}[h(1 + k)a(a + b + kb) - (1 + h)b(2a + kb)]. \end{cases} \tag{12}$$

Before going into the analysis of this system we shall first study some special, but interesting, cases. First we shall just recover the well-known selfsimilar evolution of the Wulff shape, and then we shall study the special cases where both h and k are zero and the case where $h > 0$, $k = 0$.

6.1. Some special cases

Case a = b. The special case $a = b$ corresponds to the evolution of the Wulff shape, in this special case the two relations do match exactly giving the same equation which reads

$$(1 - hk)a' = \frac{2}{a^3(2 + k)}[k(1 + h)a^2(2 + k) - (1 + k)a^2(2 + k)]$$

$$= \frac{2}{a}[k(1 + h) - (1 + k)] = \frac{2}{a}(hk - 1)$$

which gives the expected result $a' = -\frac{2}{a}$.

Case h = k = 0. For $h = k = 0$ system (12) reduces to

$$\begin{cases} a' &= -\frac{1}{a} - \frac{1}{b} \\ b' &= -\frac{2}{a}. \end{cases} \tag{13}$$

We can eliminate the time variable t and write a as a function of b to obtain

$$\frac{da}{db} = \frac{a + b}{2b}$$

for which we have the general solution

$$a(t) = b(t) + C\sqrt{b(t)},$$

with $C \in \mathbb{R}$. For $C > 0$ we see that the ratio $a(t)/b(t)$ approaches $+\infty$ as b tends to zero, which means that the Wulff shape deforms under the flow and shrinks to a *flat point*. This is interesting, as it shows that the selfsimilar evolution of the Wulff shape is actually unstable. For $C < 0$ we see that a becomes zero when $\sqrt{b} = -C > 0$, so that Σ now shrinks to a segment of strictly positive length.

Figure 2 shows the evolution starting from different values of $a(0)$ and $b(0)$ in the (b, a) phase plane. The zoom in the right of the picture clearly shows that the ratio $\frac{a}{b}$ degenerates while the surface shrinks.

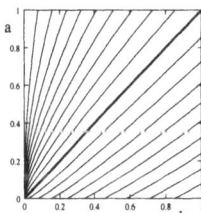
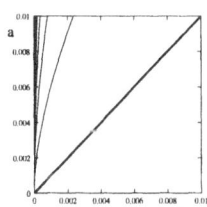

FIGURE 2. Phase diagram for $k = 0.$ and $h = 0.$

Case k = 0, h > 0. Substituting $k = 0$ in (12) we get

$$\begin{cases} a' &= -\frac{1}{a} - \frac{1}{b} \\ b' &= -\frac{2 + h}{a} + \frac{h}{b} \end{cases} \tag{14}$$

$$\frac{da}{db} = \frac{a + b}{2b + bh - ah}.$$

Setting $z = \frac{a}{b}$ we have $\frac{da}{db} = b\frac{dz}{db} + z$, whence

$$b\frac{dz}{db} = \frac{(1-z)(1-hz)}{2+h-zh}$$

which can be solved by separation of variables:

$$\int \frac{2+h-zh}{(1-z)(1-hz)}dz = \log b + \log C$$

and after some computations we get the final expression

$$a(t) = b(t) + C(b(t) - ha(t))^{\frac{1+h}{2}}.$$

Figure 3 shows the evolution starting from different values of $a(0)$ and $b(0)$ in the (b, a) phase plane. The zoom in the right of the picture clearly shows that the ratio $\frac{a}{b}$ degenerates while the surface shrinks if $a(0) < b(0)$, while it converges to some stable value if $a(0) > b(0)$.

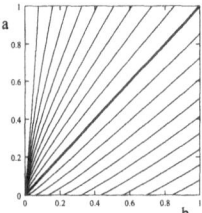
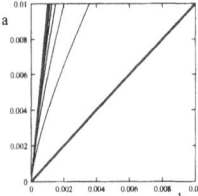

FIGURE 3. Phase diagram for $k = 0.$ and $h = 0.1$.

6.2. General case

We first look for selfsimilar solutions of system (12), suppose that $b = pa$ for some $p > 0$, we get

$$(1-hk)a' = \frac{2}{pa(2+pk)}[pk(1+h)(2+pk) - (1+k)(1+p+pk)]$$

$$= \frac{2}{pa(2+pk)}[p^2k^2 + 2phk + p^2hk^2 - 1 - k - p - pk^2]$$

$$(1-hk)a' = \frac{2}{p^2a(2+pk)}[h(1+k)(1+p+pk) - p(1+h)(2+pk)]$$

$$= \frac{2}{p^2a(2+pk)}[h + phk + hk + phk + phk^2 - 2p - p^2k - ph - p^2hk].$$

Equating the right-hand sides we get

$$p^3k^2(1+h) + p^2(k+3hk-1-k^2) + p(1+h-k-2hk-hk^2) - h(1+k) = 0$$

which is zero as expected if $p = 1$. This allows us to factor out the $(p-1)$ term to get

$$p^2k^2(h+1) + p(k^2h+3kh+k-1) + h(k+1) = 0.$$

For small values of h and k the two solutions p_1 and p_2 satisfy $0 < p_1 < 1 < p_2$ with $p_1 \approx h$ and $p_2 \approx \frac{1}{k^2}$. The explicit solution of system (12) can be computed, but that entails a lengthy computation of rational integrals, we prefer instead to show in Figure 4 the numerical results for $h = 0.1$ and $k = 0.3$, with a number of trajectories for $(b(t), a(t))$ and the stable selfsimilar evolutions in bold. As can be seen the case $p = 1$ (evolution of the Wulff shape) corresponds to an unstable evolution, while the selfsimilar evolutions with $p = p_1$ and $p = p_2$ are both stable. The dotted line in the left figure, positioned slightly above the lower selfsimilar stable evolution corresponds to the *calibrability limit*, i.e. below that line the calibrability criterion for the evolution is no longer valid, hence the subsequent evolution cannot be described by our simple two-parameter model.

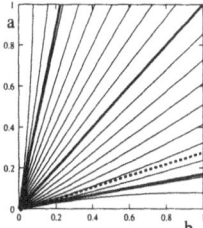 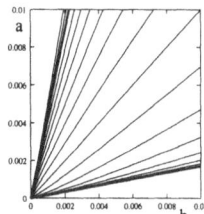

FIGURE 4. Phase diagram for $k = 0.3$ and $h = 0.1$.

An obvious difference with the case $k = 0$ is the absence of the phenomenon of shrinking to a segment of nonzero length when starting from $a(0) < b(0)$.

7. Second example: prismatic shape

The computations shown in the previous section for a cylindrically symmetric crystalline anisotropy apply without any change if we start from an hexagonal prism with all faces at distance 1 from the origin (snowflake anysotropy) instead of the cylinder and successively add an hexagonal pyramid of height h at the top and bottom faces, and finally enlarge the hexagon at $y = 0$ such that its sides have distance $1 + k$ from the origin. A vertical cross-section, both of the Wulff shape and of the evolving surface $\Sigma(t)$, would look exactly the same as in Figure 1. More precisely each point (r, y) on the polygonals of Figure 1 tells us that the cross-section at height y of the Wulff shape (or $\Sigma(t)$) is a regular polygon (with $m > 4$ sides) having all sides at distance r from the y-axis.

The number of sides m of the polygons can be different than 6, but the case $m = 4$ could give rise to difficulties which we do not want to explore here.

The only computation of Sections 4 and 5 which has to be reconsidered is that of the curvature κ_{LAT}.

Assuming that the lateral faces are calibrable, the curvature can be easily computed by enforcing the formula for the crystalline curvature obtained by J. Taylor in [1], where we only need to compute the values of weight coefficients α_i,

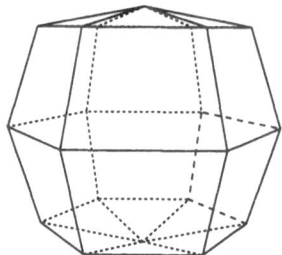

FIGURE 5. Wulff shape for the prism case.

$i = 1, \ldots, 4$ in the formula

$$\kappa_{\mathrm{LAT}} = \frac{\sum_{i=1}^{4} \alpha_i l_i}{|\Sigma_{\mathrm{LAT}}|} \tag{15}$$

where l_i denote the length of the four sides of Σ_{LAT} which is a trapezium lying above the $y = 0$ plane having the top base and bottom base of length respectively l_1 and l_3 and lateral side of length $l_2 = l_4$. Let l denote the length of the side of a m-sided polygon with unit apothem, then a straightforward computation gives

$$\alpha_1 = \frac{1-k}{\sqrt{1+k^2}}, \quad \alpha_2 = \alpha_4 = \frac{l(1+k)}{\sqrt{1+k^2}\sqrt{4+4k^2+l^2k^2}}, \quad \alpha_3 = \frac{k+k^2}{\sqrt{1+k^2}}.$$

We can now compute the four lengths l_i and the area $|\Sigma_{\mathrm{LAT}}|$ as follows

$$
\begin{aligned}
l_1 &= al \\
l_2 &= l_4 = \frac{1}{2}b\sqrt{4+4k^2+l^2k^2} \\
l_3 &= (a+kb)l \\
|\Sigma_{\mathrm{LAT}}| &= \frac{1}{2}(2a+kb)lb\sqrt{1+k^2}
\end{aligned}
\tag{16}
$$

The case $a = b = 1$ corresponds to the case where Σ is the Wulff shape itself, where we actually know that $\kappa_{\mathrm{LAT}} = 2$; we indeed recover this value by enforcing (15).

We now apply formula (15) to get

$$
\begin{aligned}
\kappa_{\mathrm{LAT}} &= \frac{2}{(2a+kb)(1+k^2)lb}\left[(1-k)al + bl(1+k) + (k+k^2)(a+kb)l\right] \\
&= \frac{2}{(2a+kb)(1+k^2)b}\left[(a+b+kb)(1+k^2)\right] = 2\frac{a+b+kb}{b(2a+kb)}
\end{aligned}
$$

which is exactly the same result obtained for the rotationally symmetric Wulff shape (6).

As a result the evolution law (given by $a(t)$ and $b(t)$) is exactly the same computed in Section 6 as long as the faces remain calibrable.

8. Calibrability analysis

Regarding the top faces (which are trangles similar to the corresponding top triangle of the Wulff shape) the computation of the Cahn-Hoffmann vector field is directly obtained from the Cahn-Hoffmann vector field at the boundary of the Wulff shape, whence the calibrability of those faces.

On the contrary, the calibrability of the lateral faces is not guaranteed. Since they are convex, we can enforce the calibrability criterion established in [6, Theorem 8.1]: for all the sides $i = 1, \ldots 4$ of the face we require $\frac{\tilde{l}_i}{l_i} \leq \kappa_{\mathrm{LAT}}$, where $\tilde{l}_i$ is the length of the side of the Wulff shape parallel to the side i of the face, and can be simply obtained from (16) by setting $a = b = 1$.

We thus obtain three conditions for the calibrability of the face. The first, related to the top side, is given by $\frac{1}{a} \leq \kappa_{\mathrm{LAT}}$ and reads exactly the same as condition (8) for the rotationally symmetric case.

The second condition is related to the bottom side (number 3), for which we need $\frac{1+k}{a+kb} \leq \kappa_{\mathrm{LAT}}$. A straightforward computation shows that this condition is always satisfied.

The third condition relates to the lateral side (number 2 and 4), for which we need $\frac{1}{b} \leq \kappa_{\mathrm{LAT}}$, and this condition also is always satisfied.

9. Conclusions

We tried here just to point out how the anisotropic evolution by mean curvature in 3D can lose some of the well-established properties enjoyed in 2D or in the isotropic case in 3D.

This is quite clear for the cylindrical or prismatic anisotropy ($h = k = 0$). The effort of studying ways to generalize those simple situations ($h, k > 0$) is motivated as a hint to suggest that instability could perhaps appear also for regular and strictly convex anisotropies. Our study is of course in no way complete in such respect and other tools are needed to answer the question whether instability of the evolution of the Wulff shape is possible or not for regular and strictly convex anisotropies.

We also recall that the stability results in 2D are strictly related to uniqueness of selfsimilarly shrinking convex curves, and to our knowledge such uniqueness result is restricted to symmetric anisotropies.

The evolutions computed for $h > 0$ also show the existence of convex selfsimilar evolutions different from the Wulff shape itself; if also $k > 0$ than there are two such evolutions. These evolutions are stable and attract evolutions starting from a perturbed Wulff shape.

An interesting byproduct of our analysis also shows that for $h = k = 0$ (cylindrical Wulff shape), starting with $b(0) > a(0)$ we have that the surface shrinks to a segment of finite length, while if $b(0) < a(0)$ the surface shrinks to a point approaching the shape of a flat disk.

The evolutions obtained with $k > 0$ are not completely determined, if $b(0) >$ $a(0)$, since at some critical time $t^* > 0$ the calibrability conditions will no longer be satisfied. After such critical time we expect some bending of faces, and the determination of the subsequent evolution is more delicate.

All numerical simulations and figures of this paper where obtained using free software (octave for the numerical simulations, gnuplot and xfig for the Figures) on a *GNU/Linux* computer system.

References

[1] F. Almgren and J.E. Taylor: *Flat flow is motion by crystalline curvature for curves with crystalline energies*, J. Differential Geom., vol 42, 1995, 1–22

[2] Bellettini and M. Novaga: *Approximation and comparison for nonsmooth anisotropic motion by mean curvature in R^N*, Math. Models Methods Appl. Sci., vol 10, 2000, 1–10

[3] G. Bellettini, M. Novaga, and M. Paolini: *Facet-breaking for three dimensional crystals evolving by mean curvature*, Interfaces and Free Boundaries, vol 1, 1999, 39–55

[4] M. Paolini: *On a crystalline variational problem, part I: first variation and global L^∞-regularity*, Arch. Ration. Mech. Anal., vol 3, 2001, 165–191

[5] M. Paolini: *On a crystalline variational problem, part II: BV-regularity and structure of minimizers on facets*, Arch. Ration. Mech. Anal., vol 3, 2001, 193–217

[6] M. Paolini: *Characterization of facet-breaking for nonsmooth mean curvature flow in the convex case*, Interfaces and Free Boundaries, vol 3, 2001, 415–446

[7] G. Bellettini and M. Paolini: *Some results on minimal barriers in the sense of De Giorgi applied to driven motion by mean curvature*, Rend. Accad. Naz. Sci. XL Mem. Mat. (5), vol 19, 1995, 43–67

[8] M. Paolini: *Quasi-optimal error estimates for the mean curvature flow with a forcing term*, Differential Integral Equations, vol 8, 1995, 735–752

[9] M. Paolini: *Anisotropic motion by mean curvature in the context of Finsler geometry*, Hokkaido Math. J., vol 25, 1996, 537–566

[10] Y.G. Chen, Y. Giga, and S. Goto: *Uniqueness and existence of viscosity solutions of generalized mean curvature flow equations*, J. Differential Geom., vol 33, 1991, 749–78

[11] E. De Giorgi: *New conjectures on flow by mean curvature*, Nonlinear variational problems and partial differential equations (Isola d'Elba, 1990), Longman Sci. Tech., Harlow, 1995, 120–128

[12] P. De Mottoni and M. Schatzman: *Geometrical evolution of developed interfaces*, Trans. Amer. Math. Soc., vol 347, 1995, 1533–1589

[13] K. Deckelnick and G. Dziuk: *Error estimates for a semi-implicit fully discrete finite element scheme for the mean curvature flow of graphs*, Interfaces and Free Boundaries, vol 2, 2000, 341–359

[14] L.C. Evans and J. Spruck: *Motion of level sets by mean curvature. I*, J. Differential Geom., vol 33, 1991, 635–681

[15] F. Fierro and M. Paolini: *Numerical evidence of fattening for the mean curvature flow*, Math. Models Methods Appl. Sci., vol 6, 1996, 793–813

[16] M.E. Gage: *Evolving plane curves by curvature in relative geometries*, Duke Math. J., vol 72, 1993, 441–466

[17] M.A. Grayson: *The heat equation shrinks embedded plane curves to round points*, J. Differential Geom., vol 26, 1987, 285–314

[18] G. Huisken: *Local and global behaviour of hypersurfaces moving by mean curvature*, Differential geometry: partial differential equations on manifolds (Los Angeles, CA, 1990), Amer. Math. Soc., Providence, RI, 1993

[19] R.H. Nochetto, M. Paolini, and C. Verdi: *Quadratic rate of convergence for curvature dependent smooth interfaces: a simple proof*, Appl. Math. Lett., vol 7, 1994, 59–63

[20] M. Paolini: *A dynamic mesh algorithm for curvature dependent evolving interfaces*, J. Comput. Phys., vol 123, 1996, 296–310

[21] R.H. Nochetto and C. Verdi: *Combined effect of explicit time-stepping and quadrature for curvature driven flows*, Numer. Math., vol 74, 1996, 105–136

[22] M. Paolini: *Convergence past singularities for a fully discrete approximation of curvature driven interfaces*, SIAM J. Numer. Anal., vol 34, 1997, 490–512

[23] A. Stancu: *Asymptotic behavior of solutions to a crystalline flow*, Hokkaido Math. J., vol 27, 1998, 303–320

[24] J.E. Taylor: *Constructions and conjectures in crystalline nondifferential geometry*, Differential Geometry, Pitman Monographs and Surveys in Pure and Applied Math. 52 (B. Lawson and K. Tenenblat, eds.), Longman Scientific and Technical, 1991, 321–336

Maurizio Paolini
Franco Pasquarelli
Dipartimento di Matematica e Fisica
Università Cattolica di Brescia
I-25121 Brescia
Italy
E-mail address: m.paolini@dmf.unicatt.it
E-mail address: f.pasquarelli@dmf.unicatt.it

Progress in Nonlinear Differential Equations
and Their Applications, Vol. 51, 155–170

Dimension Reduction in Continuum Mechanics

Danilo Percivale

Abstract. This lecture is devoted to some progress of recent ideas in Γ-convergence with varying domains for dimensional reduction problems. More precisely, we consider, as a sequence of functionals, the stored energy of a thin isotropic linear elastic body possibly damaged at meso-scale, whose cross section or whose thickness is very small. Making this parameter go to zero, by using a suitable notion of convergence of minimizers, we identify a "limit" functional which can be viewed as the elastic energy of the bar after being reduced to a one-dimensional structure or, alternatively, as the elastic-plastic energy of a thin plate.

1. Introduction

A long standing problem in continuum mechanics is to derive the constitutive equations of thin structures from those of three-dimensional bodies both in the elastic and in the elastic-plastic case. The difficulty consists in describing the behaviour of these structures with sufficient precision without solving a 3D variational problem. The idea we present here is, roughly speaking, the following: given for instance a 3D beam of small cross section (ε will denote its radius), assume that $\mathcal{F}^\varepsilon$ describes its total energy (i.e. stored energy + load energy). We try to find 1D-functionals (in the sense that they depend on functions defined on the axis of the beam) $\mathcal{F}_1, \mathcal{F}_2, \mathcal{F}_3 \ldots$ in such a way that

$$\min \mathcal{F}^\varepsilon = \varepsilon \min \mathcal{F}_1 + \varepsilon^2 \min \mathcal{J}_2 + \varepsilon^3 \min \mathcal{J}_3 + \cdots + o(\varepsilon^n)$$

and a function $\overline{\mathbf{u}}^\varepsilon$ defined in the whole beam such that

$$\mathcal{F}^\varepsilon(\overline{\mathbf{u}}^\varepsilon) = \varepsilon \min \mathcal{F}_1 + \varepsilon^2 \min \mathcal{F}_2 + \varepsilon^3 \min \mathcal{F}_3 + \cdots + o(\varepsilon^n)$$

which therefore may be considered as an *approximate minimizer* of $\mathcal{F}^\varepsilon$. In the sequel we deal with three particular situations, namely

(i) St. Venant's Beam.

(ii) Thin Notched Beams.

(iii) The Elastic-Plastic Plate.

1.1. St. Venant's beam

The behaviour of a linearly elastic homogeneous and isotropic beam bounded by a cylindrical surface (the lateral surface of the beam), by a pair of planes normal to the lateral surface (the bases of the cylinder) and subjected both to body forces and to surface tractions is well known in the literature as the St. Venant's problem or some of its generalizations. The problem consists in minimizing, in a class of suitable competing functions, a functional of the type

$$F(\mathbf{u}) = \int_\Omega \left\{ \mu |e(\mathbf{u})|^2 + \frac{\lambda}{2} |\operatorname{tr} e(\mathbf{u})|^2 - \mathbf{g} \cdot \mathbf{u} \right\} dx - \int_{\Gamma_0 \cup \Gamma_1} \mathbf{h} \cdot \mathbf{u}$$

where $\mathbf{u}$ is the displacement field, Ω the region occupied by the beam, Γ_0, Γ_1 the two bases of the cylinder, $\mathbf{g}, \mathbf{h}$ the body and surface load respectively.

The classical approach is the St. Venant's semi-inverse method: when body forces vanish and in presence of constant surface loads one may obtain explicit solutions [39], [52]. On the other hand, if we consider a beam whose length is large in comparison with the radius of the cross section then the distribution of the surface force field (if supposed smooth) over the ends "has no appreciable influence on the character of the solution in portions of the beam sufficiently far removed from the ends (see [52])" and therefore we may assume that $\mathbf{h}$ is constant on each base.

This property – known as the St. Venant's principle [35], [39] – holds even if $\mathbf{g} \neq 0$ but in general it is impossible to obtain the exact solution to the St. Venant's problem. The method developed in [45] (see Section 1 below for details) permits to construct an approximate minimizer by solving 1D differential equations.

1.2. Thin notched beams

Structural behaviour of slender elastic beams is altered by geometrical discontinuities resulting from notches. The main hindrance in describing the effect of such discontinuities in one-dimensional models is due to the fact that the classical Saint-Venant's principle, starting-point for most of beam theories, cannot possibly be true in these cases, see Toupin [57]. So, several mechanical models have been proposed in the technical literature to include the effect of a notch in one-dimensional systems, and an account of them can be found in Petroski [48]. A recurring assumption in modelling a notch is that its effects essentially have a local character, inducing a beam stiffness reduction in a region close to the discontinuity, and rapidly decaying far from the notch, [22]. Usually, physical experiments are carried out in order to characterize quantitatively the stiffness perturbation produced by real notches of various widths and depths, [53].

A different approach to notch modelization in linear elastic straight beams subject to longitudinal loads is presented here. The basic idea comes from the papers by Acerbi, Buttazzo and Percivale [1], Anzellotti, Baldo and Percivale [6] and Percivale [45], where it is shown that classical one-dimensional models of slender beams with smooth longitudinal profile can be obtained as variational limit of the corresponding three-dimensional elasticity problems when the radius of the cross-section goes to zero. By adopting a flow of ideas which is typical of the Γ-convergence of functionals defined on measures, in this paper we extend some of the above results to include the occurrence of abrupt geometrical changes on the longitudinal profile of the beam such as those induced by a notch.

It turns out that the three-dimensional elasticity problem for a notched rod under axial loads converges to a class of one-dimensional models with a localized compliance when the beam becomes very slender and when the ratio between the depth and the width of each notch vanishes in a suitable way. Roughly speaking, in the limit family the classical one-dimensional model for axial deformations holds for the two segments of beam adjacent to the notch, and the notch is represented by means of a translational elastic spring located at the damaged cross-section. This macroscopic description of localized discontinuities in beams is well known in the technical field and it is usually assumed by engineers for studying damage detection problems via dynamic methods, [2], [43]. As a matter of fact, experiments show that the analytical model with a localized compliance describe accurately the behaviour of a notched beam under axial loads as well as the classical model do for a homogeneous beam. Asymptotic behaviour of energy functionals in dimension reduction problems has been studied within this setting by Anzellotti, Baldo and Percivale in [6] for beams with circular cross section having constant radius, and by Percivale in [45] for beams with arbitrary but constant cross section. However, we extend here the analysis to the more general situation when the cross section varies along the axis, both for the scalar case, in conductivity, and for the vector one of elasticity. Referring to the latter, in this brief presentation, for its more relevant interest, we let some cylindrical portion of the beam to shrink, either radially or axially, in accordance with powers of ε of higher order with respect to the complement. We then expect that the limit model should provide discontinuous displacements, whose jumps are due to the arising of pointwise concentrations of elastic rigidities, such as springs located in the thinner regions.

1.3. The elastic-plastic plate

The asymptotic behaviour of variational integrals depending on vector-valued functions with free discontinuities may describe damaged thin structures by mean of dimension reduction in the framework of Γ-convergence.

A simplified model of such stored energy is given by a functional of the following kind

$$\int_\Sigma |\nabla^2 w|^2 \, d\mathcal{L}^2 \; + \; \mathcal{H}^1(S_{Dw}) \; + \; \int_{S_{Dw}} |[Dw]| \, d\mathcal{H}^1 \,, \tag{1}$$

where $\Sigma \subset \mathbb{R}^2$ is the reference configuration of the plate, $w \in SBH(\Sigma)$ is the scalar-valued transverse displacement, $\nabla^2 w$ is the absolutely continuous part of $D^2 w$ and S_{Dw} denotes the singular set of Dw, $\mathcal{L}$ and $\mathcal{H}$ denote, respectively, the Lebesgue and Hausdorff measure. Functional (1) has been studied by Carriero, Leaci and Tomarelli in [17], [18], [19], [20], [56].

The stored energy of elastic bodies with small cracks undergoing small deformations can be described in the frame of Special Bounded Deformation functions
([10]). We introduce here the following model of the related stored energy (see [47]):

$$\int_U |\mathcal{E}(\mathbf{v})|^2 \, d\mathcal{L}^3 + \mathcal{H}^2(J_\mathbf{v}) + \int_{J_\mathbf{v}} |[\mathbf{v}] \odot \nu_\mathbf{v}| \, d\mathcal{H}^2 \tag{2}$$

where the open set $U \subset \mathbb{R}^3$ is the reference configuration of the body, $\mathbf{v} : U \subset \mathbb{R}^3 \to \mathbb{R}^3$ is a vector field with special bounded deformation (that is to say, $\mathbf{v}$ belongs to $SBD(U)$), $\mathcal{E}(\mathbf{v})$ is the absolutely continuous part of the linear strain tensor $e(\mathbf{v}) = \text{sym} \, D\mathbf{v}$, $J_\mathbf{v}$ is the jump set of $\mathbf{v}$ (the points $\mathbf{x}$ where $\mathbf{v}$ has two different one-sided Lebesgue limits $\mathbf{v}^+(\mathbf{x}), \mathbf{v}^-(\mathbf{x})$ with respect to a suitable direction $\nu_\mathbf{v}(\mathbf{x})$), while $[\mathbf{v}] = \mathbf{v}^+ - \mathbf{v}^-$ and $\odot$ denotes the symmetric tensor product.

The first term in (2) represents the elastic energy in undamaged regions, the second one is a surface energy (area of material surfaces where damage occurs [37], the third one describes a weak resistance of the material to compression or crack opening and is related to Barenblatt model of damage [8]. The last term allows us to deal with nontrivial loads, even without artificial confinement of the body.

We study a variational relationship between the functionals (1) and (2) by showing that the elastic-perfectly plastic energy (1) of the plate is the variational limit (as $\varepsilon \to 0^+$) of the fracture-elastic energy of a three-dimensional body with thickness of order ε described as follows, where appropriate weights are given to the various terms in (0.2): we set $\Sigma^\varepsilon = \Sigma \times (-\varepsilon, \varepsilon)$ and

$$\frac{3}{2} \int_{\Sigma^\varepsilon} |\mathcal{E}(\mathbf{v})|^2 \, d\mathcal{L}^3 + \frac{\varepsilon^2}{2} \mathcal{H}^2(J_\mathbf{v}) + \varepsilon \int_{J_\mathbf{v}} |[\mathbf{v}] \odot \nu_\mathbf{v}| \, d\mathcal{H}^2 . \tag{3}$$

In [47] Percivale and Tomarelli prove that minimizers of energy (0.1) are limit of minimizers of (3) as the thickness 2ε goes to zero, and that ε^{-3} re-scaled energy of (2) converges to (3), without a priori assuming any formal asymptotic expansion of (3) minimizers.

2. Thin beams

We present in this section some results about the behaviour of linearly elastic homogeneous and isotropic beams and the modelization of notches in linear elastic straight beams subject to axial loads.

2.1. St. Venant's beam

Let $S \subset \mathbb{R}^2$ be an open bounded set with Lipschitz boundary; we denote with x_2, x_3 the variables in S and we assume that $meas(S) = 1$ and

$$\int_S x_2 x_3 \, dx_2 dx_3 = \int_S x_2 \, dx_2 dx_3 = \int_S x_3 dx_2 dx_3 = 0 \tag{4}$$

For every $\varepsilon > 0$ we will consider the set

$$\Sigma^\varepsilon = \{(\varepsilon x_2, \varepsilon x_3) : (x_2, x_3) \in S\}, \tag{5}$$

the cylinder $\Omega^\varepsilon = (0,1) \times \Sigma^\varepsilon$ and we will denote by $\Omega^\varepsilon(x_1)$ the x_1-cross section of Ω^ε that is

$$\Omega^\varepsilon(x_1) = \{(x_2, x_3) \in \Sigma^\varepsilon : (x_1, x_2, x_3) \in \Omega^\varepsilon\}.$$

Let us suppose that Ω^ε is occupied by a linearly elastic isotropic material, let $\mathbf{u}$ be the displacement field over Ω^ε and

$$e(\mathbf{u}) = \frac{1}{2}\left(\nabla \mathbf{u}^t + \nabla \mathbf{u}\right)$$

the strain field associated with $\mathbf{u}$. The strain energy $\mathcal{W}^\varepsilon(\mathbf{u})$ corresponding to a displacement field $\mathbf{u} \in H^1(\Omega^\varepsilon, \mathbb{R}^3)$ is given by

$$\mathcal{W}^\varepsilon(\mathbf{u}) = \int_{\Omega^\varepsilon} \mu\,|e(u)|^2 + \frac{\lambda}{2}|tr(e(u))|^2\,dx \tag{6}$$

where μ and λ are the Lamé moduli with $\mu > 0$, $2\mu + 3\lambda > 0$.

We suppose that the cylinder Ω^ε is subject to a body force field $\mathbf{g}^\varepsilon$ which is assumed of the type

$$\mathbf{g}^\varepsilon = g_1(x_1)\mathbf{e}_1 + \varepsilon^2(g_2(x_1)\mathbf{e}_2 + g_3(x_1)\mathbf{e}_3) + g_4(x_1)(x_3\mathbf{e}_2 - x_2\mathbf{e}_3) \tag{7}$$

where $(\mathbf{e}_1, \mathbf{e}_2, \mathbf{e}_3)$ is the canonical basis in $\mathbb{R}^3$, $g_1 \in H^1(0,1), g_2, g_3 \in L^2(0,1)$

We then define

$$\mathcal{L}^\varepsilon(\mathbf{u}) = \int_{\Omega^\varepsilon} \mathbf{g}^\varepsilon \mathbf{u}\,dx \tag{8}$$

$$\mathcal{F}^\varepsilon(\mathbf{u}) = \mathcal{W}^\varepsilon(\mathbf{u}) - \mathcal{L}^\varepsilon(\mathbf{u}) \tag{9}$$

The body force field acting on the cylinder is the superposition of two body force fields (namely, $g_1(x_1)\mathbf{e}_1 + \varepsilon^2(g_2(x_1)\mathbf{e}_2 + g_3(x_1)\mathbf{e}_3)$ and $g_4(x_1)(x_3\mathbf{e}_2 - x_2\mathbf{e}_3)$) having null momentum with respect to the x_1-axis and null resultant respectively.

We will consider the two following minimization problems

$$(\mathbf{DT}^\varepsilon) \qquad \mathrm{Min}\left\{\mathcal{F}^\varepsilon(\mathbf{u}) : \mathbf{u} = 0 \quad \text{on} \quad \{0\} \times \Sigma^\varepsilon,\, \mathbf{u} \in H^1(\Omega^\varepsilon, \mathbb{R}^3)\right\}$$

$$(\mathbf{PT}^\varepsilon) \qquad\qquad \mathrm{Min}\left\{\mathcal{F}^\varepsilon(\mathbf{u}) : \mathbf{u} \in H^1(\Omega^\varepsilon, \mathbb{R}^3)\right\}$$

It is well known that problem $\mathbf{DT}^\varepsilon$ has a unique solution while problem $\mathbf{PT}^\varepsilon$ has a solution (in general not unique) provided that

$$\int_{\Omega^\varepsilon} g_\varepsilon(x)\,dx = 0$$

In order to study properties of such minimizers, we need to introduce some auxiliary functions defined on S. We denote with $\psi \in H^1_0(S)$ the unique solution of

$$\begin{cases} \Delta\psi = -2 \\ \psi \in H^1_0(S) \end{cases}$$

which will be called the *Prandtl stress function*. To avoid technical difficulties, which are not essential in this context, we assume that S is simply connected and then it remains uniquely determined (up to constants) the *torsion function* φ defined by

$$\begin{cases} \psi_{x_3} = -\varphi_{x_2} - x_3 \\ \psi_{x_2} = \varphi_{x_3} - x_2 \,. \end{cases}$$

It is easy to see that

$$\begin{cases} \Delta\varphi = 0 \\ \frac{\partial \varphi}{\partial \mathbf{n}} = \langle(-x_3, x_2), \mathbf{n}\rangle \,. \end{cases}$$

Finally, we define the *torsional rigidity* of the cylinder as

$$\mu\tau(S) = \mu \int_S |\nabla\psi|^2 \, dx_2 dx_3 = 2\mu \int_S \psi \, dx_2 dx_3.$$

and we set for every $(x_2, x_3) \in S_\varepsilon$, $\psi_\varepsilon(x_2, x_3) = \psi(x_2/\varepsilon, x_3/\varepsilon)$ and $\varphi_\varepsilon(x_2, x_3) = \varphi(x_2/\varepsilon, x_3/\varepsilon)$. We define for every $v_1, v_4 \in H^1(0,1)$, $v_2, v_3 \in H^2(0,1)$ the functionals

$$\mathcal{F}_1(v_1) = \int_0^1 \left(\frac{1}{2} E(\lambda, \mu)|\dot{v}_1|^2 - g_1 v_1 \right) dx_1$$

$$\mathcal{F}_2(v_2, v_3, v_4) = \int_0^1 \left(E(\lambda, \mu)\{I_2(S)|\ddot{v}_2|^2 + I_3(S)|\ddot{v}_3|^2\} + \frac{\mu\tau(S)}{2}|\dot{v}_4|^2 \right) dx_1$$

$$- \int_0^1 \{I_2 p_2 \dot{v}_2 + I_3 p_3 \dot{v}_3\} + \int_0^1 \{I_0 g_4 v_4 + g_2 v_2 + g_3 v_3\} \, dx_1$$

where for every $\alpha = 2, 3$

$$I_\alpha(S) = \int_S x_\alpha^2 \, dx_2 dx_3, \qquad I_0(S) = \int_S (x_2^2 + x_3^2) \, dx_2 dx_3$$

and $E(\lambda, \mu) = \mu(2\mu + 3\lambda)/(\mu + \lambda)$.

We set now

$$q^\varepsilon(\mathbf{u}) = \left(\fint_{S^\varepsilon} \mathbf{u} \, dx_2 dx_3, (\tau(S))^{-1} \int_{S^\varepsilon} (\psi_{x_3}^\varepsilon u_2 - \psi_{x_2}^\varepsilon u_3) \, dx_2 dx_3 \right),$$

and we state our main result

Theorem 2.1. *Let* $\mathbf{u}^\varepsilon$ *be a minimizer of* $\mathbf{DT}^\varepsilon$. *Then* $q^\varepsilon(\mathbf{u}^\varepsilon) \rightharpoonup (u_1, u_2, u_3, u_4)$ *in* $H^1(0,1)$ *and*

(i) u_1 *is the unique minimizer of*

$$\min\{\mathcal{F}_1(v) : v \in H^1(0,1), v(0) = 0\} = m_1$$

and

$$\varepsilon^{-2}\mathcal{F}^\varepsilon(\mathbf{u}^\varepsilon) \to \mathcal{F}_1(u_1) = m_1;$$

(ii) $u^* = (u_2, u_3, u_4)$ *is the unique minimizer of the problem*

$$\min\{\mathcal{F}_2(v^*) : v_\alpha \in H^2(0,1), v_4 \in H^1(0,1), v^*(0) = 0 = \dot{v}_\alpha(0)\} = m_2$$

and

$$\varepsilon^{-4} \left(\mathcal{F}^\varepsilon(\mathbf{u}^\varepsilon) - \varepsilon^2 m_1 \right) \to m_2.$$

2.2. Thin notched beams

Let
$$\Sigma^\varepsilon = \{(x_1, x_2, x_3) \in \mathbb{R}^3 \ : \ x_1 \in [a, b), \ (x_2, x_3) \in S_\varepsilon(x_1)\}$$
where the cross sections $S_\varepsilon(x_1)$ are two-dimensional open balls with Lipschitz continuous radii satisfying $C\varepsilon \geq r_\varepsilon(x_1) \geq c_\varepsilon > 0$ for suitable contants C and c_ε. We consider the functionals $\mathcal{F}_\varepsilon$ defined on $W^{1,2}(\Sigma_\varepsilon; \mathbb{R}^3)$ by
$$\mathcal{F}_\varepsilon(u) = \int_{\Sigma^\varepsilon} f(e(u)) \, dx + \chi_a(u)$$
where $\chi_a(u)$ is a function which takes the value 0 if $u = 0$ on $S_\varepsilon(a)$ and $+\infty$ otherwise,
$$e(u) = (e_{ij}(u))_{i,j=1,2,3}, \quad e_{ij} = \frac{u_{i,j} + u_{j,i}}{2}$$
is the linearized strain tensor associated to the displacement vector u and $f(e)$ is the deformation energy density. Moreover we shall make the assumption that the body is linearly elastic and isotropic, so that
$$f(e) = \mu|e|^2 + \frac{\lambda}{2}|\mathrm{tr}(e)|^2$$
where $\lambda, \mu > 0$ are the Lamé constants of the material and $\mu > 0$, $2\mu + 3\lambda > 0$. We assume that the structure is loaded only by axial forces which are expressed by a vector field g of the form $(g(x_1), 0, 0)$, so that their scaled potential energies are given by
$$\mathcal{L}_\varepsilon(u) = \frac{1}{\varepsilon^2} \int_{\Sigma_\varepsilon} g(x_1)u_1(x) \, dx.$$
The stored energy of the one-dimensional limit structure will be the functional
$$\mathcal{F}(v) = \begin{cases} \dfrac{\pi E}{2}\left(\displaystyle\int_a^b \left|\frac{dv'}{dm}\right|^2 dm + \frac{|v(a^+)|^2}{m(\{a\})}\right) & \text{if } v \in BV(a, b) \text{ and } v' \ll m \\ +\infty & \text{elsewhere in } L^1(a, b), \end{cases}$$
where $E = \mu\dfrac{3\lambda + 2\mu}{\lambda + \mu}$ is the Young's modulus of the material. Finally, the limit potential energy is given by
$$\mathcal{L}(v) = \int_a^b g(x_1)v_1(x_1) \, dx_1.$$
By setting
$$v_i(x_1, x_2, x_3) = u_i\big(x_1, r_\varepsilon(x_1)x_2, r_\varepsilon(x_1)x_3\big), \quad i = 1, 2, 3$$
we obtain
$$v_{i,1} = u_{i,1} + \frac{\dot{r}_\varepsilon}{r_\varepsilon}x_2 v_{i,2} + \frac{\dot{r}_\varepsilon}{r_\varepsilon}x_3 v_{i,3}$$
$$v_{i,2} = u_{i,2}r_\varepsilon$$
$$v_{i,3} = u_{i,3}r_\varepsilon,$$

the transformed functional $\mathcal{F}_\varepsilon^\Omega : W^{1,2}(\Omega; \mathbb{R}^3) \to \mathbb{R}$ is given by

$$
\mathcal{F}_\varepsilon^\Omega(v) = \frac{\mu}{\varepsilon^2} \int_\Omega \Big[|r_\varepsilon v_{1,1} - \dot{r}_\varepsilon x_2 v_{1,2} - \dot{r}_\varepsilon x_3 v_{1,3}|^2 + |v_{2,2}|^2 + |v_{3,3}|^2
$$
$$
+ \frac{1}{2}|v_{1,2} + r_\varepsilon v_{2,1} - \dot{r}_\varepsilon x_2 v_{2,2} - \dot{r}_\varepsilon x_3 v_{2,3}|^2
$$
$$
+ \frac{1}{2}|v_{1,3} + r_\varepsilon v_{3,1} - \dot{r}_\varepsilon x_2 v_{3,2} - \dot{r}_\varepsilon x_3 v_{3,3}|^2 + \frac{1}{2}\big[v_{2,3} + v_{3,2}\big]^2 \Big] dx
$$
$$
+ \frac{\lambda}{2\varepsilon^2} \int_\Omega \Big[|r_\varepsilon v_{1,1} - \dot{r}_\varepsilon x_2 v_{1,2} - \dot{r}_\varepsilon x_3 v_{1,3}|^2 + |v_{2,2}|^2 + |v_{3,3}|^2
$$
$$
+ (r_\varepsilon v_{1,1} - \dot{r}_\varepsilon x_2 v_{1,2} - \dot{r}_\varepsilon x_3 v_{1,3})(v_{2,2} + v_{3,3}) + 2 v_{2,2} v_{3,3} \Big] dx + \chi_a(u).
$$

while
$$
\mathcal{L}_\varepsilon^\Omega(v) = \frac{1}{\varepsilon^2} \int_\Omega r_\varepsilon^2(x_1) g(x_1) v_1(x) \, dx.
$$

Let $q : W^{1,2}(\Omega) \to W^{1,2}(a,b)$, defined by

$$
q(v) = \frac{1}{\pi} \int_{S_1} v(x_1, x_2, x_3) \, dx_2 dx_3.
$$

Theorem 2.2. *Let us assume that*

$$
\frac{\varepsilon^2}{r_\varepsilon^2(x_1)} dx_1 \to m \quad weakly* \ in \ \mathcal{M}[a,b],
$$

$$
\frac{\varepsilon^2}{r_\varepsilon^2(x_1)} \dot{r}_\varepsilon^2(x_1) dx_1 \to 0 \quad weakly* \ in \ \mathcal{M}[a,b].
$$

If v^ε minimizes $\mathcal{F}_\varepsilon^\Omega + \mathcal{L}_\varepsilon^\Omega$ then

(i) *if (ε_n) is a sequence such that $\lim_{n\to\infty} \varepsilon_n = 0$ and if $q(v_1^{\varepsilon_n})$ converges to some v in $L^1(a,b)$ then v is a minimizer of $\mathcal{F} + \mathcal{L}$ and $\lim_{n\to\infty} \big[\mathcal{F}_{\varepsilon_n}^\Omega(v^{\varepsilon_n}) + \mathcal{L}_{\varepsilon_n}^\Omega(v^{\varepsilon_n}) \big] = \mathcal{F}(v) + \mathcal{L}(v)$;*

(ii) *there is a sequence ε_n such that $\lim_{n\to\infty} \varepsilon_n = 0$ and a minimizer v of $\mathcal{F} + \mathcal{L}$ such that $q(v_1^{\varepsilon_n})$ converges to v weakly* in $BV(a,b)$, hence strongly in $L^1(a,b)$.*

3. The elastic plastic plate

We assume from now on that $\Omega \subset \mathbb{R}^2$ is a bounded connected Lipschitz open set, that $\overline{\Omega} \subset \Sigma = (-L, L)^2$ and we set, for every $\varepsilon \in (0,1]$,

$$
\Omega^\varepsilon = \Omega \times (-\varepsilon, \varepsilon), \quad \Sigma^\varepsilon = \Sigma \times (-\varepsilon, \varepsilon), \quad \Gamma = \partial\Omega, \quad \Gamma^\varepsilon = \Gamma \times (-\varepsilon, \varepsilon). \quad (10)
$$

We denote with $\mathbf{x} = (x, y)$ the variables on Ω, with z the variable on $(-\varepsilon, \varepsilon)$ and with t the re-scaled vertical variable variable on $(-1, 1)$. For every vector field $\mathbf{W} : \Omega^1$ (respectively Ω^ε) $\to \mathbb{R}^3$ we denote with $\mathbf{w} = (w_1, w_2)$ its two horizontal components, with w_3 its vertical one and for every $\mathbf{X} \in \Omega^1$ (resp. Ω^ε) we set $\mathbf{X} = (\mathbf{x}, t)$ and $\mathbf{X} = (\mathbf{x}, z)$ with $\mathbf{x} = (x, y) \in \Sigma$, $t \in (-1, 1)$ and $z \in (-\varepsilon, \varepsilon)$ respectively.

Ω^ε will be the reference configuration of a thick elastic-plastic body which we will assume loaded by a dead force field $\mathbf{F}^\varepsilon$, *not necessarily perpendicular to the middle surface* Ω, such that

$$\mathbf{F}^\varepsilon(x, y) = \frac{\varepsilon^2}{2} \mathbf{G}(x, y) \qquad \mathbf{G} = (g_1, g_2, g) \in L^p(\Omega, \mathbb{R}^3) \quad , p \in [3, +\infty] \quad (11)$$

Without re-labeling we denote by $\mathbf{F}^\varepsilon$ and $\mathbf{G}$ also their trivial extensions on Σ^ε (say spt F^ε, spt $G \subset \overline{\Omega^\varepsilon}$).

Given λ, μ (the Lamé constants) such that $\mu > 0$, $2\mu + 3\lambda > 0$, $\delta > 0$, $\gamma > 0$, we assume that the stored energy due to a displacement $\mathbf{V} \in SBD(\Sigma^\varepsilon)$ is

$$\mathcal{G}^\varepsilon(\mathbf{V}) = \int_{\Sigma^\varepsilon} \left(\mu |\mathcal{E}(\mathbf{V})|^2 + \frac{\lambda}{2} |\operatorname{Tr} \mathcal{E}(\mathbf{V})|^2 \right) d\mathbf{X} + \int_{J_\mathbf{V}} \theta^\varepsilon([\mathbf{V}], \nu_\mathbf{V}) \, d\mathcal{H}^2(\mathbf{X}) \quad (12)$$

where

$$\theta^\varepsilon(\eta, \xi) = \varepsilon^2 \delta |\xi| + \varepsilon \gamma |\eta \odot \xi| . \quad (13)$$

Now we introduce the load energy and the total energy associated to the displacement field $\mathbf{V}$:

$$\mathcal{L}^\varepsilon(\mathbf{V}) = \int_{\Sigma^\varepsilon} \mathbf{F}^\varepsilon \cdot \mathbf{V} \, d\mathbf{X} = \frac{\varepsilon^2}{2} \int_{\Sigma^\varepsilon} \mathbf{G} \cdot \mathbf{V} \, d\mathbf{X} \quad (14)$$

$$\mathcal{F}^\varepsilon(\mathbf{V}) = \mathcal{G}^\varepsilon(\mathbf{V}) - \mathcal{L}^\varepsilon(\mathbf{V}) \quad (15)$$

and we state the weak formulation for the Dirichlet problem (see also [10])

$(\mathbf{LP}^\varepsilon)$ $\qquad\qquad \min \{ \mathcal{F}^\varepsilon(\mathbf{V}) : \mathbf{V} \in SBD(\Sigma^\varepsilon), \text{ spt } \mathbf{V} \subset \overline{\Omega^\varepsilon} \} .$

We summarize here some facts about the space *SBD* mentioned above. The space of vector fields with bounded deformation (*BD*) has been introduced to deal with variational problems in perfect plasticity (see [54],[55]):

$$BD(U) = \left\{ \mathbf{v} \in L^1(U, \mathbb{R}^n) : \ e(\mathbf{v}) := \frac{1}{2} \left(D\mathbf{v} + (D\mathbf{v})^T \right) \in \mathcal{M}(U, M_{n,n}) \right\}$$

$$\|\mathbf{v}\|_{BD(U)} = \|\mathbf{v}\|_{L^1(U)} + \int_U |e(\mathbf{v})| .$$

$BD(U)$ is the dual of a separable Banach space. For any $\mathbf{v} \in BD(U)$ we define the jump set $J_\mathbf{v}$ which is the subset of the singular set where $\mathbf{v}$ has one-sided approximate limits with respect to a suitable direction $\nu_\mathbf{v}$ "normal" to $J_\mathbf{v}$. $J_\mathbf{v}$ is called the *jump set* of $\mathbf{v}$ and plays a role analogous to the singular set $S_\mathbf{v}$ in the theory of BV functions (see [4]).

We notice that for $\mathbf{v} \in BV(U, \mathbb{R}^k)$, the set $S_\mathbf{v} \setminus J_\mathbf{v}$ is $\mathcal{H}^{n-1}$ negligible, while it is not known whether the same property holds in $BD(U)$. Moreover for every $\mathbf{v} \in BD(U)$ the linear strain tensor $e(\mathbf{v})$ has the following decomposition

$$e(\mathbf{v}) = e^a(\mathbf{v}) + e^s(\mathbf{v}) = \mathcal{E}(\mathbf{v})dx + e^j(\mathbf{v}) + e^c(\mathbf{v})$$

where $e^a(\mathbf{v}) = \mathcal{E}(\mathbf{v})dx$ and $e^s(\mathbf{v})$ are respectively the absolutely continuous and the singular part of $e(\mathbf{v})$ with respect to $\mathcal{L}^n$; $e^j(\mathbf{v})$, $e^c(\mathbf{v})$ are respectively the restriction of $e^s(\mathbf{v})$ to $J_\mathbf{v}$ and the restriction of e^s to its complement (say the *jump* and *Cantor part* of $e(\mathbf{v})$). From now on we denote by $\mathrm{div}\,\mathbf{v} = \mathrm{Tr}\,\mathcal{E}(\mathbf{v})$ the absolutely continuous part of the distributional divergence of $\mathbf{v}$.

$\mathcal{E}(\mathbf{v})$ can be interpreted as an approximate symmetric differential (Th. 4.3 [4]) that is $\mathcal{L}^n$a.e. $x \in U$

$$\lim_{\varrho \to 0^+} \frac{1}{\varrho^n} \int_{B_{\varrho x}} \frac{\left|\left(\mathbf{v}(\mathbf{y}) - \mathbf{v}(\mathbf{x}) - \mathcal{E}(\mathbf{v}) \cdot (\mathbf{y} - \mathbf{x})\right) \cdot (\mathbf{y} - \mathbf{x})\right|}{|\mathbf{y} - \mathbf{x}|^2}\, d\mathbf{y} \;=\; 0\;,$$

$J_\mathbf{v}$ is a Borel set with null Lebesgue measure and is countably $(\mathcal{H}^{n-1}, n-1)$ rectifiable (see [4],Prop. 3.5), and there are $\mathbf{n_v} = \nu_\mathbf{v}(x) \in \partial B_1$, $\mathbf{v}^+(x), \mathbf{v}^-(x)$ (respectively geometric measure theory normal, outer and inner trace in the ν direction) $\mathcal{H}^{n-1}$ a.e. in $J_\mathbf{v}$, s.t.

$$e^j(\mathbf{v}) \;=\; (\mathbf{v}^+ - \mathbf{v}^-) \odot \nu_\mathbf{v}\, \mathcal{H}^{n-1} \llcorner J_\mathbf{v}\;,$$

and the jump part $e^j(\mathbf{v})$ can be represented on every Borel set B by the formula

$$e^j(\mathbf{v})(B) \;=\; \int_{B \cap J_\mathbf{v}} [\mathbf{v}] \odot \nu_\mathbf{v}\, d\mathcal{H}^{n-1}\;, \qquad \text{where } [\mathbf{v}] := \mathbf{v}^+ - \mathbf{v}^-.$$

We say that a function $\mathbf{v} \in BD$ belongs to SBD when its linear strain tensor $e(\mathbf{v})$ has vanishing Cantor part. If $\mathcal{R}$ denotes the set of *rigid displacements* (the affine maps of type $A \cdot x + b$ where $A \in M_{n,n}$ is skew-symmetric and $b \in \mathbb{R}^n$), then ([54] Proposition 2.2, 2.3 p. 155 and Theorem 3.1 [4]) for every bounded connected open set U with Lipschitz boundary, and every continuous linear map $R : BD(U) \to \mathcal{R}$ which fixes the elements of $\mathcal{R}$, there is a constant $c_1 = c_1(U, R)$ such that

$$\|\mathbf{v} - R(\mathbf{v})\|_{L^{n/(n-1)}(U)} \;\leq\; c_1(U, R)\, |e(\mathbf{v})|(U) \qquad \forall \mathbf{v} \in BD(U)\;.$$

If ψ is a continuous semi-norm on $BD(U)$ and a norm on $\mathcal{R}$, then $\psi(v) + \int_U |e(\mathbf{v})|$ is a norm on $BD(U)$ equivalent to $\|\cdot\|_{BD(U)}$.

3.1. A Poincaré-type inequality

The asymptotic behaviour, when ε goes to zero, of the constant in a Poincaré inequality for a cylinder Ω^ε and in general for an ε-fattened open subset of $\mathbb{R}^{n-1}$, cannot be estimated simply by scaling: when the body is supposed to be fixed at the lateral boundary the results stated below give a sharp estimate of the Poincaré constant and constitutes a crucial step in proving the main result of this section.

Theorem 3.1. *([47]) For $n \geq 2$ let $\Omega \subset \mathbb{R}^{n-1}$ be a non empty bounded connected Lipschitz open set and*

$$\overline{\Omega} \subset (-L, L)^{n-1} \stackrel{\mathrm{def}}{=} \Sigma \tag{16}$$

and set

$$\Gamma = \partial\Omega\;, \quad \Omega^\varepsilon = \Omega \times (-\varepsilon, \varepsilon)\;, \Sigma^\varepsilon = \Sigma \times (-\varepsilon, \varepsilon)\;, \Gamma^\varepsilon = \partial\Omega \times (-\varepsilon, \varepsilon)$$

for every $\varepsilon \in (0,1]$. *Then there exists a constant* $C_\Omega = C(\Omega, L, n) > 0$ *independent of* ε *such that*

$$\|\mathbf{v}\|_{L^{n/(n-1)}(\Omega^\varepsilon)} \leq C_\Omega \varepsilon^{-1-1/n}|e(\mathbf{v})|(\Sigma^\varepsilon) \leq C_\Omega \varepsilon^{-1-1/n}\left(|e(\mathbf{v})|(\Omega^\varepsilon) + \int_{\Gamma^\varepsilon} |\mathbf{v}|\, d\mathcal{H}^{n-1}\right)$$

for every $\varepsilon \in (0,1]$ *and for every* $\mathbf{v} \in BD(\Sigma^\varepsilon)$ *such that* spt $\mathbf{v} \subset \overline{\Omega^\varepsilon}$.

The proof of the previous result depends essentially on the following lemma which is precisely the same of Theorem 3.1 for piecewise-rigid displacements.

Proposition 3.2. *There is* $c_\Omega = c(\Omega, L, n) > 0$ *such that for every* $\mathbf{w} = \sum_k \mathbf{w}_k \mathbf{1}_{Q_k}$, *with* spt $\mathbf{w}$ *disjoint from* $\partial\Sigma \times (-\varepsilon, \varepsilon)$ *and* $\mathbf{w}_k$ *in* $\mathcal{R}(Q_k)$ $\forall k$, *the following estimate holds*

$$\|\mathbf{w}\|_{L^{n/(n-1)}(\Omega^\varepsilon)} \leq c_\Omega \varepsilon^{-1-1/n}|e(\mathbf{w})|(\Sigma^\varepsilon)$$

for every ε *such that* $0 < \varepsilon \leq 1$.

Remark 3.3. *The rate* $\varepsilon^{-1-1/n}$ *is optimal. In fact, if* $n \geq 2$, *and for* $\mathbf{x} \in \Sigma$, $z \in (-\varepsilon, \varepsilon)$, *we choose* $\varphi \in C^2(\Sigma)$ *with* $\varphi \not\equiv 0$, spt $\varphi \subset \Omega$ *and set*

$$\mathbf{v}(\mathbf{x}, z) = (-zD\varphi(\mathbf{x}), \varphi(\mathbf{x}))^T$$

Clearly, spt $\mathbf{v} \subset \Omega^\varepsilon$, $\mathbf{v} \in SBD(\Sigma^\varepsilon)$, $J_{\mathbf{v}^\varepsilon} = \emptyset$ *and*

$$\mathcal{E}(\mathbf{v}) = \begin{pmatrix} -zD^2\varphi & 0 \\ 0^T & 0 \end{pmatrix}$$

Therefore,

$$\int_{\Sigma^\varepsilon} |e(\mathbf{v})| = \int_{\Sigma^\varepsilon} |\mathcal{E}(\mathbf{v})|\, d\mathbf{x} = \varepsilon^2 \|D^2\varphi\|_{L^1(\Sigma)}$$

while, as $\varepsilon \to 0_+$,

$$\left(\int_{\Sigma^\varepsilon} |\mathbf{v}|^{\frac{n}{n-1}}\right)^{\frac{n-1}{n}} \sim \varepsilon^{\frac{n-1}{n}} 2^{\frac{n-1}{n}} \|\varphi\|_{L^s(\Sigma)}$$

which shows that the blow-up estimate is optimal.

3.2. Main results

From now on we will assume $n = 3$ and greek indices vary in the set $\{1, 2\}$, and the roman indices in $\{1, 2, 3\}$, and we will denote with $e_\mathbf{x}$ and $\mathcal{E}_\mathbf{x}$ the 2×2 tensor field whose components coincide with the horizontal components of e and $\mathcal{E}$ respectively. For every $\mathbf{V} \in SBD(\Sigma^\varepsilon)$ we define a vector field $\mathbf{U}$ in Σ^1 by a suitable re-scaling, and a family of functionals E^ε to be evaluated on $\mathbf{U}$ (so that we can refer to a single fixed domain Σ^1 instead of varying domains Σ^ε).

$$\mathbf{U}(\mathbf{x}, t) = \left(\varepsilon^{-1}\mathbf{v}(\mathbf{x}, \varepsilon t), v_3(\mathbf{x}, \varepsilon t)\right), \qquad \mathbf{x} \in \Sigma,\ z \in (-\varepsilon, \varepsilon),\ t \in (-1, 1),$$

$$E^\varepsilon(\mathbf{U}) = \varepsilon^{-3}\mathcal{F}^\varepsilon(\mathbf{V})$$

Actually such $\mathbf{U}$ belongs to $SBD(\Sigma^1)$ (see [47]).

We fix $\mathbf{V}^\varepsilon \in \operatorname{argmin} \mathcal{F}^\varepsilon$ and $\mathbf{U}^\varepsilon(\mathbf{x}, t) = (\varepsilon^{-1}\mathbf{v}^\varepsilon(\mathbf{x}, \varepsilon t), v_3^\varepsilon(\mathbf{x}, \varepsilon t))$. Then $\mathbf{U}^\varepsilon \in \operatorname{argmin} E^\varepsilon$ and spt $\mathbf{U}^\varepsilon \subset \overline{\Omega^\varepsilon}$.

We study the asymptotic behaviour of the family $\mathbf{U}^\varepsilon$ as $\varepsilon \to 0_+$ with respect to the strong convergence in $L^1(\Sigma^1)$ (shortly denoted by σ in the sequel). In order to describe the variational limit of functionals E^ε it is useful to recall the notion of Γ convergence (see [26], [25]).

Definition 3.4. *Let $\varepsilon > 0$, (S, σ) a complete metric space and $I, I^\varepsilon : S \to \mathbb{R} \cup \{+\infty\}$ a family of functionals. We say that*

$$\Gamma(\sigma^-) \lim_{\varepsilon \to 0^+} I^\varepsilon(s) \;=\; I(s)$$

if and only if the two following conditions are satisfied

i) $\forall s \in S \qquad \forall \varepsilon_n \to 0^+ \qquad \forall s_n \xrightarrow{\sigma} s \qquad \liminf_{n \to \infty} I^{\varepsilon_n}(s_n) \geq I(s),$

ii) $\forall s \in S \qquad \forall \varepsilon_n \to 0^+ \qquad \exists s_n \xrightarrow{\sigma} s \qquad \limsup_{n \to \infty} I^{\varepsilon_n}(s_n) = I(s).$

The most important consequence of Γ-convergence is convergence of minimizers.

Theorem 3.5. *([26], Cor.2.4) Assume $I \;=\; \Gamma(\sigma^-) \lim\limits_{\varepsilon \to 0^+} I^\varepsilon$. If $s^\varepsilon \in \operatorname{argmin} I^\varepsilon$ and $s_\varepsilon \xrightarrow{\sigma} s$, then*

$$I^\varepsilon(s^\varepsilon) \;\to\; I(s) \;=\; \min\{I(s) : \; s \in S\} \,.$$

Now we can state the two main results

Theorem 3.6. *Assume (10)–(16) and the safe load condition*

$$\|\mathbf{G}\|_{L^p(\Omega)} \;<\; \frac{\sqrt[3]{4}\,\gamma\,|\Sigma|^{\frac{3-p}{3p}}}{C_\Omega} \,. \tag{17}$$

Set

$$\mathcal{A} = \{\mathbf{U} \in SBD(\Sigma^1) : \mathbf{U} = (\mathbf{u}, u_3),\, u_3 = u_3(x,y),\, \operatorname{spt} u_3 \subset \overline{\Omega},\, \mathbf{u}(x,y,t) = \zeta - tDu_3\}$$

$$E^0(\mathbf{U}) \;=\; \int_\Sigma \left\{ \frac{2}{3}\mu \left((|\nabla^2 u_3|^2 + \frac{\lambda}{\lambda + 2\mu}|\Delta^a u_3|^2 \right) + \right.$$

$$\left. + 2\mu \left(|\mathcal{E}_x(\zeta)|^2 + \frac{\lambda}{\lambda + 2\mu}|\operatorname{div}\zeta|^2 \right) \right\} dx dy +$$

$$+ \int_{J_\mathbf{U}} (\delta + \gamma|[\mathbf{U}] \odot \nu_\mathbf{U}| \, d\mathcal{H}^2(x,y,t) - \int_\Sigma g u_3 \, dx dy \qquad \text{if } \mathbf{U} \in \mathcal{A}$$

$E^0(\mathbf{U}) \;=\; +\infty$ *otherwise. Then*

$$\Gamma(L^1(\Sigma^1, \mathbb{R}^3)^-) \lim_{\varepsilon \to 0^+} E^\varepsilon(\mathbf{U}) \;=\; E^0(\mathbf{U}).$$

Remark 3.7. *It is worth noticing that if $\mathbf{U} = (\mathbf{u}, u_3) \in \mathcal{A}$ then $\mathbf{u}(.,t) \in SBD(\Sigma)$ for a.e. $t \in (-1,1)$ and hence $\zeta \in SBD(\Sigma)$, $\operatorname{spt}\zeta \subset \overline{\Omega}$, $\operatorname{spt} u_3 \subset \overline{\Omega}$, $\nu_\mathbf{U}$ is a horizontal vector, say $(\nu_\mathbf{U})_3 \equiv 0$ $\mathcal{H}^2$ a.e. on $J_\mathbf{U}$ and $u_3 \in SBH(\Sigma)$, since*

$$e_x(\mathbf{U}) = e(\mathbf{u}) = e(\zeta) - t\,e(Du_3) = e(\zeta) - tD^2 u_3 \in \mathcal{M}(\Sigma) \qquad \text{a.e. } t$$

whence $e(\zeta)$ and $D^2 u_3$ belong to $\mathcal{M}$ and, since $e_x(\mathbf{U})$ has no Cantor part, $u_3 \in SBH, \zeta \in SBD$.

In particular, if $\mathbf{U} = (\mathbf{u}, u_3) \in \mathcal{A}$ *and* $E^0(\mathbf{U}) < +\infty$, *then* $\nabla^2 u_3 = \nabla Du_3 \in L^2$ *and* $\mathcal{E}(\zeta) \in L^2$.

In addition to Theorem 3.4. it is possible to prove the following result concerning convergence of minimizers of $(\mathbf{LP}^\varepsilon)$.

Theorem 3.8. *Assume (10)–(17). Then, for every* $\varepsilon \in (0,1]$ *and for every* $\mathbf{V}^\varepsilon$ *in* $\operatorname{argmin}(\mathbf{LP}^\varepsilon)$, *by setting*

$$\mathbf{U}^\varepsilon(\mathbf{x}, t) = (\mathbf{u}^\varepsilon, u_3^\varepsilon) = (\varepsilon^{-1}\mathbf{v}^\varepsilon(\mathbf{x}, \varepsilon t), v_3^\varepsilon(\mathbf{x}, \varepsilon t)) = (\varepsilon^{-1}\mathbf{v}^\varepsilon(\mathbf{x}, \zeta), v_3^\varepsilon(\mathbf{x}, \zeta)),$$

for $\mathbf{x} \in \Sigma$, $z \in (-\varepsilon, \varepsilon)$, $t \in (-1, 1)$, *we have, up to subsequences,*

$$(\mathbf{u}^\varepsilon, u_3^\varepsilon) \xrightarrow{\quad w^* - SBD(\Sigma) \quad} (-t\, Du_3\,,\, u_3)$$

for a suitable u_3 *minimizer of*

$$(\mathbf{LP}) \qquad\qquad \min\left\{ \mathcal{G}^0(w)\ :\ w \in SBH(\Sigma)\ ,\ \operatorname{spt} w \subset \overline{\Omega} \right\}$$

where

$$
\begin{aligned}
\mathcal{G}^0(w) &= \frac{2}{3}\mu \int_\Sigma \left(|\nabla^2 w|^2 + \frac{\lambda}{\lambda + 2\mu}|\Delta^a w|^2 \right) d\mathbf{x} + 2\delta\, \mathcal{H}^1(S_{Dw}) \\
&\quad + \gamma \int_{S_{Dw}} |[Dw]|\, d\mathcal{H}^1(\mathbf{x}) - \int_\Sigma gw\, d\mathbf{x}.
\end{aligned}
$$

References

[1] E. Acerbi, G. Buttazzo and D. Percivale, *Thin inclusion in nonlinear elasticity,* J. Reine Angew. Math., 386 (1988), 99–115.

[2] R.D. Adams, P. Cawley, C.J. Pye, and B.J. Stone, *A vibration technique for nondestructively assessing the integrity of structures,* Journal Mechanical Engineering Science **20** (1978), no. 2, 93–100.

[3] L. Ambrosio *Existence theory for a new class of variational problems,* Arch. Rational Mech. Anal., 111 (1990), 291–322.

[4] L. Ambrosio, A. Coscia & G. Dal Maso, *Fine properties of functions with bounded deformation,* Arch. Rat. Mech. Anal., 139, 3 (1997), 201–238.

[5] L. Ambrosio & A. Braides, *Energies in SBV and variational models in fracture mechanics,* Homogeneization and Appl. to Material sciences, (Nice, 1995), 9 GAKUTO Int.Ser.Math.Sci Appl.

[6] G. Anzellotti, S. Baldo & D. Percivale, *Dimension reduction in variational problems, asymptotic development in* Γ-*convergence and thin structures in elasticity* Asymptotic Anal., 9, (1994) 61–100.

[7] C. Baiocchi, G. Buttazzo, F. Gastaldi & F. Tomarelli, *General existence theorems for unilateral problems in continuum mechanics,* Arch. Rat. Mech. Anal., **100**, 2 (1988), 149–189.

[8] G.I. Barenblatt, *The formation of equilibrium cracks during brittle fracture, general ideas and hypotheses. Axially symmetric cracks,* Appl. Math. Mech. (PMM), 23, (1959), 622–636.

[9] K. Bhattacharya & A. Braides, *Thin films with many small cracks*, Prepr. SISSA, **36** (1999), 1–16.

[10] G. Bellettini, A. Coscia & G. Dal Maso, *Compactness and lower semi-continuity properties in SBD(Ω)*, Math.Z., 228 (1998), 337–351.

[11] G. Bouchitté, *Représentation intégrale de fonctionnelles convexes sur un espace de mesures. II. Cas de l'épi-convergence.*, Ann. Univ. Ferrara Sez. VII (N.S.), **33** (1987), 113–156.

[12] A. Braides, G. Dal Maso & A. Garroni, *Variational formulation of softening phenomena in fracture mechanics: the one-dimensional case*, Arch. Rat. Mech. Anal., 146, (1999) 23–58.

[13] A. Braides & V. Chiadò-Piat, *Integral representation results for functionals defined on SBV(Ω, ℝ^m)*, J. Math. Pures Appl., 75 (1996), 595–626.

[14] A. Braides & I. Fonseca, *Brittle thin films*, Prepr. SISSA **37**, (1999), 1–26.

[15] A. Braides, I. Fonseca & G. Francfort, *3D-2D asymptotic analysis for inhomogeneous thin films*, Prepr. SISSA, **71** (1999).

[16] G. Buttazzo and L. Freddi, *Functionals defined on measures and applications to non equi-uniformly elliptic problems*, Ann. Mat. Pura Appl. (4) **159** (1991), 133–149.

[17] M. Carriero, A. Leaci & F. Tomarelli, *Plastic free discontinuities and special bounded hessian*, C. R. Acad. Sci. Paris, 314 (1992), 595–600.

[18] M. Carriero, A. Leaci & F. Tomarelli, *Special Bounded Hessian and elastic-plastic plate*, Rend. Accad. Naz. delle Scienze (dei XL), (109) XV (1992), 223–258.

[19] M. Carriero, A. Leaci & F. Tomarelli, *Strong solution for an Elastic Plastic Plate*, Calc. Var., 2 (1994), 219–240.

[20] M. Carriero, A. Leaci & F. Tomarelli, *Free gradient discontinuities*, in: Buttazzo, Bouchitté, Suquet Eds, "Calculus of Variations, Homogeneization and Continuum Mechanics", World Scientific, Singapore, 1994, 131–147.

[21] V. Casarino & D. Percivale, *A variational model for non linear elastic plates* J. Convex Anal., 3 (1996) 221–243.

[22] S. Christides and A.D.S. Barr, *One-dimensional theory of cracked Bernoulli-Euler beams*, International Journal of the Mechanical Sciences, **26** (1984), no. 11-12, 639–648.

[23] P.G.Ciarlet, *Mathematical Elasticity, vol II: Theory of Plates*, Studies in Math. and its Appl., North-Holland, (1997).

[24] F. Colombo & F. Tomarelli, *Boundary value problems and obstacle problems for an elastic body with free cracks*, to appear.

[25] G. Dal Maso, *An Introduction to Gamma Convergence* Birkhäuser, PNLDE 8 (1993).

[26] E. De Giorgi & T. Franzoni, *Su un tipo di convergenza variazionale*, Atti Accad. Naz. Lincei Rend. Cl. Sci. Mat., 58 (1975), 842–850.

[27] E. De Giorgi, *Γ-convergenza e G-convergenza*, Boll. Un. Mat. Ital., 5 14-A (1977) 213–220.

[28] E. De Giorgi, *Free discontinuity problems in calculus of variations*, Frontiers in Pure & Applied Mathematics, R. Dautray ed., North-Holland, Amsterdam 1991, 55–61.

[29] E. De Giorgi & L. Ambrosio, *Un nuovo tipo di funzionale del Calcolo delle Variazioni*, Atti Accad. Naz. Lincei, Rend. Cl. Sci. Fis. Mat. Natur., 82 (1988), 199–210.

[30] Demengel F., *Fonctions a hessien borné*, Ann. Inst. Fourier 34 (1984) 155–190.

[31] G.Del Piero, *One-dimensional ductile-brittle transition, yielding and structured deformations*, Prepr. Dip. Ing. Univ. Ferrara, 42, (1999).

[32] G. Del Piero & D.R. Owen, *Structured deformations of continua*, Arch. Rat. Mech. Anal., 124 1993, 99–155.

[33] G. Del Piero & L. Trushkinovsky, *A one-dimensional model for localized and distributed failure*, J.Phys., IV France, pr. 8, (1998), 95–102.

[34] H. Federer, *Geometric Measure Theory*, Springer, Berlin, 1969.

[35] G. Fichera, *Il principio di St. Venant: intuizione dell'ingegnere e rigore del matematico*, Rend. Mat. Ser VI **10** (1977) 1–24.

[36] I. Fonseca & G. Francfort, *3D-2D asymptotic analysis of optimal design problem for thin films*, J. reine angew. Math., 505 (1998), 173–202.

[37] A.A.Griffith, *The phenomenon of rupture and flow in solids*, Phyl.Trans. Roy.Soc. A, 221, (1920) 163–198.

[38] G. Kirchhoff, *Über das Gleichgewicht und die Bewegung einer elastischen Scheibe*, J. Reine Angew. Math. **40** (1850), 51–88.

[39] D. Iesan, *St. Venant's Problem*, Lecture Notes in Math., Vol 1279, Springer, 1987.

[40] L.D. Landau & E.M. Lifshitz, *Theory of elasticity*, 7 Th. Phys., (1959) Pergamon Press.

[41] J.L. Lions *Problèmes aux limites en théorie des distributions*, Acta Math. **94** (1955), 13–153.

[42] P. Marcollini and C. Sbordone, *An approach to the asymptotic behaviour of elliptic-parabolic operators*, J. Math. Pures Appl. **56** (1977), 157–182.

[43] A. Morassi, *Identification of a crack in a rod based on changes in a pair of natural frequencies*, J. Sound Vibration, To appear.

[44] A. Morassi, *A uniqueness result on crack location in vibrating rods*, Inverse Problems in Engineering **4** (1997), 231–254.

[45] D.Percivale, *Perfectly plastic plates: a variational definition*, J. Reine Angew. Math. **411** (1990), 39–60.

[46] D. Percivale & F. Tomarelli, *Scaled Korn-Poincaré inequality in BD and a model of elastic plastic cantilever*, Asympt. Analysis, **23**, (2000), 291–311.

[47] D. Percivale & F. Tomarelli, *From SBD to SBH: the elastic plastic plate*, Interfaces and Free Boundaries, **4**, (2002) 1–29.

[48] H.J. Petroski, *Comments on 'free vibrations of beams with abrupt changes in cross-section'*, J. Sound Vibration. **92** (1984), no. 1, 157–159.

[49] P. Podio-Guidugli, *Constraint and scaling methods to derive shell theory from three-dimensional elasticity*, Riv.Math.Univ.Parma, **16** (1990), 72–83.

[50] G. Savarè & F. Tomarelli, *Superposition and chain rule for Bounded Hessian Functions*, Advances in Math., **140** (1998), 237–281.

[51] M.A. Save & C.E. Massonet, *Plastic Analysis and design of plates, shells and disks*, North-Holland Ser. in Appl. Math. and Mech. (1972).

[52] I.S. Sokolnikoff, *Mathematical Theory of Elasticity*, Mc Graw-Hill, 1956.

[53] W.T. Springer, K.L. Lawrence, and T.J. Lawley, *The effect of a symmetric disconti-nuity on adjacent material in a longitudinally vibrating uniform beam*, Experimental Mechanics (1987), 168–171.

[54] R. Temam, *Problèmes Mathematiques en Plasticité*, Gauthier-Vllars, (1983), Paris.

[55] R. Temam & G. Strang, *Functions of bounded deformation*, Arch. Rat. Mech. Anal., 75 (1980), 7–21.

[56] F. Tomarelli, *Special Bounded Hessian and partial regularity of equilibrium for a plastic plate*, in: G. Buttazzo, G.P. Galdi, L. Zanghirati (Eds.), "Devel. in PDE and Applications to Mathematical Physics", Plenum Press, N. Y., 1992, 235–240.

[57] R.A. Toupin, *Saint-Venant's principle*, Arch. Rational Mech. Anal. 18 (1965), no. 2, 83–96.

[58] L. Truskinowsky, *Fracture as a phase transition*, Preprint.

[59] W.P. Ziemer, *Weakly Differentiable Functions*, Springer, N.Y., 1988.

Danilo Percivale
Dipartmento di Metodi e Modelli Matematici
Università di Genova
P.le J.F.Kennedy Pad. D
I-16129 Genova
Italia
E-mail address: percivale@dimet.unige.it

Progress in Nonlinear Differential Equations
and Their Applications, Vol. 51, 171–182
© 2002 Birkhäuser Verlag Basel/Switzerland

Reaction-Diffusion Equations and Learning

Jayant Shah

Abstract. A reaction-diffusion equation is formulated for synthesizing textures. The parameters of the equation are estimated from examplars by the maximum likelihood principle.

1. Introduction

An effective approach for modelling many problems in Computer Vision is provided by variational calculus. In this approach, an energy functional is formulated containing a linear combination of terms (also called potentials in the probabilistic context), each of which is a nonlinear transformation of the output of a linear filter such as the gradient or the laplacian of the smoothed image intensity (see §2). Diffusion equations are derived by gradient descent to find solutions minimizing the energy functional. An energy functional induces a Gibbs probability distribution on the space of all images. A question then arises whether this probability distribution agrees with what is observed in nature. To address this question, Zhu, Wu and Mumford [7] applied the entropy minimax principle to the problem of texture synthesis and estimated the probability distribution. Remarkably, the basic form of the component potentials is qualitatively the same as some of the ad-hoc potentials already in use such as the Blake–Zisserman [2] and the Perona-Malik potentials [3] and the potentials used by Geman and Graffigne for stochastic modelling of textures [4]. It is also similar to the "edge-strength" function encountered in the segmentation problem [5] and the sigmoid function used in neural nets (see §3).

A brief description of the entropy minimax method of Zhu, Wu and Mumford is given in §2. The entropy minimax principle is a powerful principle first formulated by Christensen [8], [9], [10] in the context of pattern recognition and statistical inference. Zhu, Wu and Mumford independently formulated it for application to Computer Vision. The problem is to model the probability distribution on the feature space or the space of images. Since entropy is inversely related to the amount of information in the model, its maximization ensures that the model contains no more information than what is present in the observed sample. Minimization of entropy is used to find the model that captures the maximum amount of information from the sample.

This work was supported by NIH Grant I-R01-NS34189-04 and NSF Grant DMS-9531293

To apply the entropy minimax principle, both Christensen and Zhu et al. partition the feature space and approximate the probability distribution by a piecewise constant function. The problem is thus reduced to a problem in parametric statistics, albeit with a greatly increased number of parameters. The principle of maximum entropy is used to estimate the probability distribution corresponding to a given partition. Christensen uses the principle of minimum entropy to find the optimum partitioning of each feature space which consumes most of the computational effort. Zhu, Wu and Mumford propose the principle of minimum entropy to implement feature pursuit so that features are introduced in the order of their importance and usually the first few features suffice to represent the sample adequately. The main computation in their formulation is in estimating the parameters by the principle of maximum entropy which amounts to maximizing the log-likelihood function by gradient ascent. The computationally intensive part is concerned with the synthesis of a sample image by the Gibbs sampler from the current estimates of the parameters during each update. By fitting smooth curves to the piecewise constant potentials, Zhu and Mumford discovered that all have the same basic form depending on three parameters, its center, scale, and its rate of growth. The equation of gradient descent applied to the corresponding energy functional with smoothed potentials has the form of reaction-diffusion equation. The objective of this paper is to present a deterministic formulation in which the Gibbs sampler is replaced by a deterministic reaction-diffusion equation. Its parameters are estimated by means of the maximum likelihood principle.

2. Entropy minimax

What follows is a brief summary of the entropy minimax formulation of Zhu, Wu and Mumford. Details may be found in [7]. In practice, the method amounts to an application of the maximum likelihood principle and feature pursuit based on residuals. Start with the Gibbs form of probability distribution:

$$p(I) = \frac{1}{Z} e^{-U(I)} \tag{1}$$

$$U(I) = \sum_{\alpha} \int_D \phi^{(\alpha)} \left(I^{(\alpha)} \right)$$

where I is an image, $I^{(\alpha)}$ is a linear transform of I, $\phi^{(\alpha)}(\xi)$ is a nonlinear function, D is the image domain and Z is the partition function. Of course, $U(I)$ is the corresponding energy functional. The problem is to estimate the potential functions $\phi^{(\alpha)}(\xi)$. Consider one of the $\phi^{(\alpha)}(\xi)$'s and to simplify the notation, denote it by ϕ, omitting the superscript. Divide the domain of ϕ into M bins. Let χ_i denote the characteristic function of the ith bin:

$$\chi_i = \begin{cases} 1 & \text{if } \xi \in i^{th} \text{ bin} \\ 0 & \text{otherwise} \end{cases} \tag{2}$$

Let χ_i denote the coordinate of the center of the ith bin. Let $|D|$ denote the area of the image domain. Then,

$$\phi(\xi) \approx \sum_{i=1}^{M} \phi(\xi_i)\chi_i(\xi) \tag{3}$$

$$\int_D \phi(\xi(\mathbf{x}))d\mathbf{x} \approx \sum_{i=1}^{M} \lambda_i f_i(\xi)$$

where $\lambda_i - |D|\phi(\zeta_i)$ and f_i is the normalized frequency:

$$f_i(\xi) = \frac{1}{|D|} \int_D \chi_i(\xi(\mathbf{x}))d\mathbf{x} \tag{4}$$

$$\approx \frac{1}{|D|} \# \left\{\mathbf{x} \in \mathbf{D} : \xi(\mathbf{x}) \in i^{th} \text{ bin}\right\}$$

The approximate probability distribution function

$$p(\Lambda, I) \approx \frac{1}{Z} \exp\left(-\sum_{\alpha}\sum_{i} \lambda_i^{(\alpha)} f_i^{(\alpha)}(I)\right) \tag{5}$$

now depends only on the parameters $\{\lambda_i^{(\alpha)}\}$. The normalized frequency $f_i^{(\alpha)}(I)$ depends only on $I^{(\alpha)}$ and not on ϕ.

The same form for the probability distribution function is derived in [7] by applying the principle of maximum entropy as follows. The problem is cast as that of maximizing the entropy

$$\int p(I) \log \frac{1}{p(I)} dI \tag{6}$$

with respect to p, subject to the constraints

$$E_p[f_i^{(\alpha)}] = f_{i,obs}^{(\alpha)} \tag{7}$$

where $E_p[\cdot]$ denotes the expected value of the feature with respect to p and $f_{i,obs}^{(\alpha)}$ is the observed mean value of the feature. Then application of the method of Lagrange multipliers shows that the probability distribution function must have form (1).

The entropy is maximized indirectly with respect to the parameters λ_i^{α} by applying the maximum likelihood principle as follows. By applying gradient ascent to the log-likelihood function, we get

$$\frac{d\lambda^{(\alpha)}}{dt} = E_{p(\Lambda)}[f_i^{(\alpha)}] - f_{i,obs}^{(\alpha)} \tag{8}$$

where Λ denotes the set of current values of the parameters λ_i^{α}. Since the log-likelihood function here is convex, there is a unique maximizing Λ and hence a unique Λ satisfying Eq. (7). Therefore the Λ maximizing the likelihood function also solves the constrained entropy maximization problem.

It is not feasible to compute the expected values in Eq. (7). However, they may be estimated using the ergodicity theorem of Geman and Geman [6]: Synthesize a typical image $I_{syn,\Lambda}$ from the distribution $p(\Lambda)$ and use $f_i^{(\alpha)}(I_{syn,\Lambda})$ as an estimate for $E_{p(\Lambda)}[f_i^{(\alpha)}]$. The main computation is now that of $I_{syn,\Lambda}$ and the transforms $I_{syn,\Lambda}^{(\alpha)}$. Computation of the Gibbs sampler is made managable by keeping the number of allowed pixel values and hence the number of local characteristics that must be computed low.

In order to reduce the number of features used and thereby the number of image transforms to be calculated, feature pursuit is used. At each stage of the feature pursuit, optimum value of Λ is found for the current set of chosen features. The next feature chosen should be the one which maximizes the reduction in entropy which is equivalent to minimizing the Kullback-Leibler distance (that is, the relative entropy) from the true probability distribution. However, it is hard to estimate entropies. Since the purpose of the gradient ascent Eq. (8) is to drive down the residual on the right-hand side to zero, a heuristic strategy is to select at each step the feature with the largest residual vector. In essence, we use the residuals to indicate the distance of $p(\Lambda)$ from the true probability distribution. Let S denote the set of features already selected. Let Λ_S denote the set of values obtained by setting λ_i^α equal to its maximum likelihood estimate if the feature α belongs to S and zero otherwise. Initially, S is empty and $I_{syn,\emptyset}$ consists of uniform noise. Define

$$d(\beta) = \sum_{i=1}^{M} |f_i^{(\beta)}(I_{syn,\Lambda_S}) - f_{i,obs}^{(\alpha)}| \tag{9}$$

Choose β such that $d(\beta)$ is maximum over the complement of the set S. The use of the L^1-norm in Eq. (9) instead of the L^2-norm or higher norms is recommended by Zhu et al. as it gave the best results in their analysis.

3. The basic potential

Zhu and Mumford derive their reaction-diffusion equation using potentials of the form:

$$\phi^{(\alpha)}(\xi) = a_\alpha \frac{(|\xi - c_\alpha|/b_\alpha)^{q_\alpha}}{1 + (|\xi - c_\alpha|/b_\alpha)^{q_\alpha}} \tag{10}$$

They arrive at this form by fitting curves to the piecewise constant potentials they found empirically. The potential is symmetric about c_α and asymptotically reaches a constant value a_α monotonically. Introduction of the shift parameter c_α is new and necessary because there is no reason why a particular feature should behave symmetrically with respect to the origin. To understand the behavior of such potentials, consider the segmentation functional:

$$E_{MS}(I, B) = \sigma^q \int_{D-B} \|\nabla I\|^q + \gamma^q |B| + \int_D |I - I_{obs}|^q \tag{11}$$

where B is the segmenting curve, $|B|$ is its length and $1 < q < \infty$. First look at the GNC algorithm of Blake and Zisserman ($q = 2$). They replace the first two terms in functional (11) by a function of $\|\nabla I\|$ which is a smoothed version of a truncated parabola and thus has essentially the same shape as given by Eq. (10) (see [11]). The Blake-Zisserman potential is symmetric about the origin. For small values of $\|\nabla I\|$, it behaves like $\sigma^2 \|\nabla I\|^2$ and the behavior is governed by diffusion. Parameter σ plays the role of $\sqrt{a_\alpha}/b_\alpha$. As $\|\nabla I\| \to \infty$, the potential saturates, monotonically approaching the constant value γ^2. The ratio γ/σ is the scale parameter corresponding to b_α in (10). It marks the transition region between diffusion and saturation. The diffusion equation of Perona and Malik may also be derived from a similar potential [11]. The trouble of course is that these approximate functionals have zero infimum and the corresponding gradient descent equations are unstable. Recently, Braides and Dal Maso have regularized the Blake and Zisserman functional by replacing $\|\nabla I\|^2$ in the functional by its average value over a neighborhood and show that the regularized functional, suitably normalized, converges to the segmentation functional (11) as the averaging radius tends to zero [12].

The form (10) is closely related also to the edge-strength function implicit in the approximation of functional (11) formulated by Ambrosio and Tortorelli [13]:

$$E_{AT}(I,v) = \sigma^q \int_D \left[(1-v)^2 \|\nabla I\|^q + \gamma^q \left(\rho \|\nabla v\|^2 + \frac{v^2}{\rho} \right) + |I - I_{obs}|^q \right] \quad (12)$$

which is valid also for the case $q = 1$. The edge-strength function v which minimizes $E_{AT}(I,v)$ is a smoothing of

$$\frac{2\rho \|\sigma \nabla I/\gamma\|^q}{1 + 2\rho \|\sigma \nabla I/\gamma\|^q} \quad (13)$$

which is identical in form to the potential (10) with zero shift.

The exponent q determines the type of diffusion which occurs when the gradient is small. The gradient flow for minimizing the q-norm of the gradient is governed by $(q-1)I_{nn} + I_{ss}$ where I_{nn} is the second derivative in the direction of ∇I and I_{ss} in the direction orthogonal to it. When $q = 2$, the diffusion is isotropic, governed by the laplacian. In the limiting case when $q = 1$, smoothing occurs by curvature-dependent evolution of the level curves of I and the gradient flow of functional (12) develops shocks [5]. As $q \to \infty$, the flow in the limit is purely in the direction of the gradient and has been analyzed by Jensen [14]. Potential (10) assumes (double) sigmoidal shape used in neural nets as $q_\alpha \to \infty$ and becomes a purely thresholding function in the limit. The scaling parameter b_α may now be thought of as a threshold for the feature in the sense of neural nets.

Potentials of the same kind in the form of edge-strength functions are also employed in the newly developed faster methods for segmenting images, notably, the method of curve evolution, which is intimately related to the segmentation functionals (11) and (12), (see [5]), and a more recent graph-theoretic method proposed by Shi and Malik [15]. The two methods are in fact closely related;

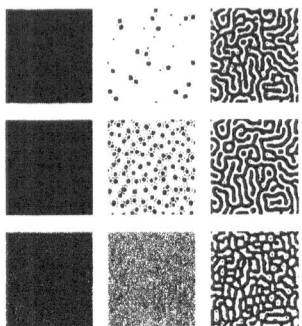

FIGURE 1. Diffusion ($a > 0$). Top row: $q = 1$; Middle row: $q = 2$;
Bottom row: $q = 4$. First column: $c = 0$; Middle column: $c = -6$;
Last column: $c = -24$

the latter may be interpreted as an approximation of the former [16]. In both approaches, the increased speed of computation is achieved basically by delinking determination of the edge-strength function from boundary detection. The edge-strength function is calculated in advance of boundary detection by means of ad-hoc potentials similar to the basic potential (10).

An important consequence of the saturation of potentials is that the functional

$$U(I) = \int_D \sum_\alpha \phi^{(\alpha)} \left(I^{(\alpha)} \right) \tag{14}$$

has not only an infimum, but also a supremum. Hence, the weights a_α may be allowed to be negative, a possibility very effectively exploited by Zhu and Mumford in [1] for synthesizing textures and removing clutter. For an illustration, consider the case of a single potential. Denote the corresponding filter by F so that the integrand in Eq. (14) is $\phi(F * I)$. The gradient flow is given by the equation

$$\frac{\partial I}{\partial t} = F_- * \phi'(F * I) \tag{15}$$

where $F_-(\mathbf{x}) = -\mathbf{F}(-\mathbf{x})$. When $a > 0$ the flow is a diffusion flow and it is reactive when $a < 0$. To illustrate the behavior, we consider the case where F smooths the image by convolving it with a Gaussian with standard deviation equal to $3/\sqrt{2}$ and then computes the laplacian of the smoothed image. Textures obtained by applying the gradient flow (15) to uniform random noise are shown in Figs. 1 and 2. The case when $a > 0$ is shown in Fig. 1 for three values of $q \geq 1$ and three values of c; the case when $a < 0$ is shown in Fig. 2. The general pattern seems to be that of blobs and stripes at various scales in both the cases. Zhu and Mumford report empirical evidence for $q < 1$. Fig. 3 shows the results of gradient flow with $q = 0.5$ and a positive as well as negative. With $a > 0$, the flow developed instabilities indicated by black square blotches in the figure. Eventually, I assumed a constant value.

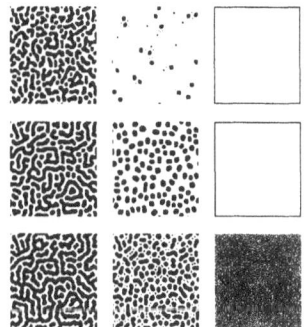

FIGURE 2. Reaction: ($a > 0$). Top row: $q = 1$; Middle row: $q = 2$; Bottom row: $q = 4$. First column: $c = 0$; Middle column: $c = 6$; Last column: $c = 24$

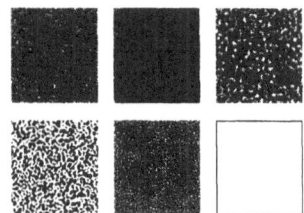

FIGURE 3. $q = 0.5$. Top row: Diffusion ($a > 0$); Bottom row: Reaction ($a < 0$). First column: $c = 0$; Middle column: $|c| = 6$; Last column: $|c| = 24$.

4. Learning textures

Each unknown potential in Functional (1) is now represented as a linear combination of fixed smooth potentials instead of piecewise constant approximation. We create such a set of potentials by shifting and scaling the "mother" potential

$$\psi(\xi) = \frac{|\xi|^q}{1 + |\xi|^q} \tag{16}$$

Function ψ is "universal" in the following sense. Let Ψ denote the set of functions of the form

$$a_o + \sum_{finite} a_i \psi\left(\frac{\xi - c_i}{b_i}\right) \tag{17}$$

Then, for each $-\infty < h < k < \infty$, the restriction of Ψ to the interval $[h, k]$ is dense in the set $C^0[h, k]$ of continuous functions on $[h, k]$ with respect to the uniform norm. This follows from a result of Leshno, Lin, Pinkus and Schoken [17] which asserts that if ψ is continuous almost everywhere and bounded on every compact set, then it is universal if and only if it is not a polynomial. Thus, for the purpose of approximating the potentials $\phi^{(\alpha)}$ by elements in Ψ, any universal function will suffice. The choice (16) is suggested by the results of Zhu and Mumford and by the form of the potentials encountered in segmentation functionals.

Let $R = \max\{\xi_{\max} - \xi_{\text{mean}}, \xi_{\text{mean}} - \xi_{\min}\}$. Define

$$\psi_{m,k}(\xi) = \psi\left(m\frac{\xi}{R} - k\right) \tag{18}$$

where m, k are integers, $m > 0$ and $|k| \leq m$. For a fixed m, range of ξ is covered by $2m+1$ potentials and the potentials get narrower and narrower as m increases. The situation is analogous to the multiscale representation of a function by wavelets except that our basis potentials are not orthogonal.

The unknown potential $\phi(\xi)$ may be approximated as

$$\phi(\xi) \approx \frac{1}{|D|}\sum_{m,k}\theta_{m,k}\psi_{m,k}(\xi) \tag{19}$$

In view of the universality of ψ, it is expected that with a sufficiently large set of fixed potentials, a single value of q will suffice for approximating the unknown potentials. Hence, q was set equal to 2 in all the experiments. Note that the main effect of q is on the type of diffusion.

The probability distribution is now given by

$$p(I, \Theta) = \frac{1}{Z}e^{-U(I,\Theta)} \tag{20}$$

where

$$U(I, \Theta) = \int_D \sum_\alpha \sum_{m,k} \frac{1}{|D|}\theta_{m,k}^{(\alpha)}\psi_{m,k}^{(\alpha)}\left(I^{(\alpha)}\right) = \sum_\alpha \sum_{m,k}\theta_{m,k}^{(\alpha)}v_{m,k}^{(\alpha)}(I) \tag{21}$$

$$\text{where} \quad v_{m,k}^{(\alpha)}(I) = \frac{1}{|D|}\int_D \psi_{m,k}^{(\alpha)}\left(I^{(\alpha)}\right)$$

Quantities $v_{m,k}^{(\alpha)}$ may be thought of as a set of nonlinear features derived from the original linear filters. The equations for gradient ascent to estimate $\theta_{m,k}^{(\alpha)}$ may be derived as before, but the synthesized image is computed by applying gradient descent to the energy functional $U(I, \Theta)$. We get a system of coupled differential equations[1]:

$$\frac{d\theta_{m,k}^{(\alpha)}}{dt} = v_{m,k}^{(\alpha)}(I_{syn}) - v_{m,k}^{(\alpha)}(I_{obs}) \tag{22}$$

$$\frac{\partial I_{syn}}{\partial t} = \sum_\alpha F_-^{(\alpha)} * \sum_{m,k}\frac{\theta_{m,k}^{(\alpha)}}{|D|}\psi_{m,k}'^{(\alpha)}\left(F^{(\alpha)} * I_{syn}\right)$$

where $F^{(\alpha)} * I_{syn}$ is the linear transform.

These equations are degenerate in the direction of the current vector Θ in the sense that the infimum of $U(I, \Theta)$ with respect to I is invariant with respect

---

[1] Similar equations have been used by Anderson and Langer [18] using the absolute value function for all α in place of $\psi_{m,k}$ and disregarding the shift and scaling parameters.

to scaling of the Θ and hence Θ must be normalized. The scaling of Θ corresponds to setting the temperature in simulated annealing. As the temperature tends to zero, the annealing method converges to the most likely image. The probability distribution in the limit is uniform over the set of the minima of $U(I, \Theta)$ and zero elsewhere [6]. Thus, as the temperature tends to zero, the probability mass becomes more and more concentrated around the global minima and the most likely images become also typical images. In this sense, the use of the most likely image for estimating expected values of $v_{m,k}^{(\alpha)}$ here is consistent.

Normalization of Θ may be done by restricting Θ to the Euclidean length of one. The first equation in (22) may then be replaced by

$$\frac{d\theta_{m,k}^{(\alpha)}}{dt} = \left[v_{m,k}^{(\alpha)}(I_{syn}) - v_{m,k}^{(\alpha)}(I_{obs}) \right]^{\perp} \tag{23}$$

where the right-hand side is the component of the residuals orthogonal to Θ.

Again, feature pursuit may be used as described in §2.

5. Experiments

The following choices were made for the three experiments described below.

The filter bank $\{F^{(\alpha)}\}$ is a subset of the filter bank used by Zhu, Wu and Mumford, consisting of 73 linear filters:

$$LG(T) = \frac{4}{\pi T^4} \left[\left(\frac{x}{T}\right)^2 + \left(\frac{y}{T}\right)^2 - 1 \right] \exp \left\{ -\left[\left(\frac{x}{T}\right)^2 + \left(\frac{y}{T}\right)^2 \right] \right\} \tag{24}$$

where $T = \sqrt{2}/2, 1, 2, 3, 4, 5, 6$ and Gabor filters

$$G\cos(T, \nu) = \frac{1}{\pi T^2} \exp\left\{ -\frac{1}{2}\left[\left(\frac{2x'}{T}\right)^2 + \left(\frac{y'}{T}\right)^2 \right] \right\} \cos\frac{2\pi x'}{T} \tag{25}$$

$$G\sin(T, \nu) = \frac{1}{\pi T^2} \exp\left\{ -\frac{1}{2}\left[\left(\frac{2x'}{T}\right)^2 + \left(\frac{y'}{T}\right)^2 \right] \right\} \sin\frac{2\pi x'}{T}$$

where $x' = x\cos\nu + y\sin\nu$, $y' = -x\sin\nu + y\cos\nu$, $T = 2, 4, 6, 8, 10, 12$, and $\nu - 0°, 30°, 60°, 90°, 120°, 150°$. The filters $G\sin(2, \cdot)$ were omitted because $G\sin(2,0)$ is identically zero at the lattice points.

The set of potentials $\psi_{m,k}$ consisted of 35 potentials with $m = 1, 2, 4, 8, 16$ and $|k| \le m/2$. The full range of centers k was not used since the centers of the unknown potentials should be near the mean values of the corresponding features.

The observed images were normalized so that the pixel values ranged from 0 to 255. The uniform noise was sampled from this range of values. The input images and the synthesized images are 128×128 pixels in size except the last input image is 79×142 pixels in size. The range of values of each feature was computed by combining the range obtained from the observed image with the range obtained from the image consisting of uniform noise.

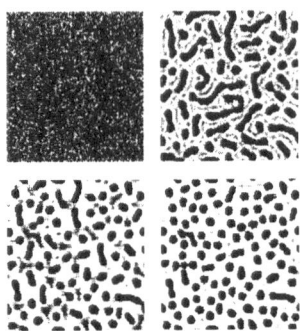

FIGURE 4. Learned texture. (a) Top left: uniform noise generated
by the system initially, (b) Top right: Texture generated after the
first filter was chosen, (c) Bottom left: synthesis using 4 filters, (d)
Bottom right: synthesis using 6 filters. (Compare with the image
in the center of Fig. 2.)

The size of the time step Δt was empirically chosen as follows. After the first
feature was chosen, Θ was set equal to the residuals with respect to that feature,
with the length adjusted to one. The time step was then chosen so that the first
update vector $\delta\Theta$ had length equal to 0.1 and this value of the time step was
maintained during all the subsequent updates. After each new feature was chosen,
Θ was updated 10 times. Each time Θ was updated, I_{syn} was computed using the
second update equation in (22). The time step Δt for updating I_{syn} was set such
that during the first update of the uniform noise, the maximum change in the pixel
values was equal to 2. The image was updated 60 times before introducing the next
feature. Note that it is not crucial to drive down the residuals to zero before a new
feature is introduced. It is sufficient to make the residuals small enough compared
to the residuals of the new feature. In order to avoid boundary effects, toroidal
topology was assumed.

In the first experiment, the system was given the image shown in the center
of Fig. 2 ($q = 2, c = 6$) as the image to be learned. The system selected six filters
in the following order: $LG(4)$, $LG(0)$, $Gcos(4, 90)$, $LG(6)$, $Gcos(4, 0)$, $Gcos(8, 90)$.
Fig. 4a shows the uniform noise with which the system begins. The synthesized
images after 1, 4 and 6 filters were selected are shown in Figs. 4b, 4c and 4d
respectively. Interestingly, although the input image was synthesized with a single
filter $LG(3)$ using Eq. (15), the system (22) chose $LG(4)$ instead as its first filter.
The values of $d(\beta)$ of the two filters are very close with the latter having a slightly
higher value.

The second test image shown in Figure 5a depicts animal fur. Figure 5b shows
the result after the system had selected 8 filters in the following order: $Gcos(2, 60)$,
$Gcos(2, 0)$, $Gcos(6, 150)$, $LG(\sqrt{2}/2)$, $Gsin(12, 0)$, $Gcos(12, 120)$, $Gcos(2, 90)$ and
$Gsin(6, 60)$.

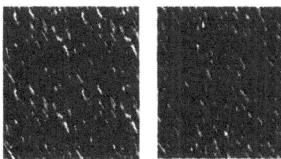

FIGURE 5. (a) Left: Texture to be learned, (b) Texture synthesized by the system.

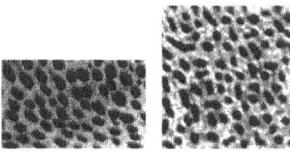

FIGURE 6. (a) Left: Texture to be learned, (b) Texture synthesized by the system.

The last experiment is shown in Figure 6. Figure 6a is the input image showing cheetah skin. Figure 6b shows the image synthesized by the system after it chose 13 filters in the following order: $LG(1)$, $Gcos(12, 150)$, $Gcos(12, 120)$, $Gsin(12, 60)$ $Gcos(10, 90)$, $Gcos(12, 0)$, $Gsin(12, 30)$, $G(6, 120)$, $LG(4)$, $Gsin(6, 30)$, $Gsin(6, 0)$, $Gsin(6, 60)$ and $Gsin(6, 150)$.

References

[1] S.C. Zhu and D. Mumford, *Prior learning and Gibbs reaction-diffusion*, IEEE Trans. PAMI **19(11)**, Nov. 1997.

[2] A. Blake and A. Zisserman: *Visual Reconstruction*, MIT Press, 1987.

[3] P. Perona and J. Malik: *Scale-space and edge detection using anisotropic diffusion*, IEEE Trans. PAMI **12(7)**, July, 1990.

[4] S. Geman and C. Graffigne: *Markov random field image models and their applications to computer vision*, International Congress of Mathematician, 1986, 1496–1517.

[5] J. Shah: *A common framework for curve evolution, segmentation and anisotropic diffusion*, IEEE Conf. on Computer Vision and Pattern Recognition, June, 1996.

[6] S. Geman and D. Geman: *Stochastic relaxation, Gibbs distribution, and the Bayesian restoration of images* ", IEEE Trans. PAMI **6**, 1984, 721–741.

[7] S.C. Zhu, Y. Wu and D. Mumford: *Filters, random fields and maximum entropy (FRAME)*, Int. J. of Computer Vision **27(2)** , March-April 1998, 1–20.

[8] R. Christensen: *A pattern discovery program for analyzing qualitative and quantitative data* ", Behavioral Science **13(5)**, September, 1968, 423–424.

[9] R. Christensen: *Entropy minimax, a non-bayesian approach to probability estimation from empirical data* ", Proc. of the 1973 International Conference on Cybernetics and Society, IEEE Systems, Man and Cybernetics Society, November 1973, 321–325.

[10] R. Christense: *Entropy minimax sourcebook, general description*, **1**, Entropy Limited, Lincoln MA, 1981.

[11] J. Shah: *Parameter estimation, multiscale representation and algorithms for energy-minimizing segmentations*, IEEE Conf. on Computer Vision and Pattern Recognition, June, 1990.

[12] A. Braides and G. Dal Maso: *Nonlocal approximation of the Mumford-Shah functional*, Calc. Var. **5**, 1997, 293–322.

[13] L. Ambrosio and V.M. Tortorelli, *On the approximation of functionals depending on jumps by quadratic, elliptic functionals*, Boll. Un. Mat. Ital, 1992.

[14] R. Jensen: *Uniqueness of Lipschitz extensions: minimizing the sup norm of the gradient*, Arch. Rational Mechanics **123**, 1993, 51–74.

[15] J. Shi and J. Malik: *Normalized cuts and image segmentation*, IEEE Conf. on Computer Vision and Pattern Recognition, June, 1997.

[16] J. Shah, Riemannian drums, curve evolution and segmentation, in *Scale-Space Theories in Computer Vision, Lecture Notes in Computer Science* **1682**, Ed: M. Nielsen, P. Johansen, O.F. Olsen and J. Weickert, Springer, 1999.

[17] M. Leshno, V.Ya. Lin, A. Pinkus and S. Schoken: *Multilayer feed-forward networks with a non-polynomial activation function can approximate any function*, Neural Networks, 1993.

[18] C.H. Anderson and W.D. Langer: *Statistical Models of Image Texture*, Technical Report, Washington University Medical School, 1997.

Jayant Shah
Mathematics Department
Northeastern University
Boston, MA
USA
E-mail address: shah@neu.edu

Contributors

EMILIO ACERBI
  Department of Mathematics
  University of Parma
  V. D'Azeglio 85/a,
  I-43100 Parma, Italy

GIOVANNI BELLETTINI
  Dipartimento di Matematica, Roma "Tor Vergata"
  Via della Ricerca Scientifica
  I-00133 Roma, Italy

GUY BOUCHITTÉ
  Département de Mathématiques
  Université de Toulon et du Var
  F-83957 La Garde Cedex, France

GIUSEPPE BUTTAZZO
  Dipartimento di Matematica, Università di Pisa
  via Buonarroti 2
  I-56127 Pisa, Italy

MICHELE CARRIERO
  Dipartimento di Matematica "Ennio De Giorgi"
  Università di Lecce
  Via Arnesano
  I-73100 Lecce, Italia

VICENT CASELLES
  Departament de Tecnologia, Universitat Pompeu-Fabra
  Passeig de Circumvalació 8
  E-08003 Barcelona, Spain

ANDREA CIANCHI
  Dipartimento di Matematica e Applicazioni per l'Architettura
  Università di Firenze
  Piazza Ghiberti 27
  I-50122 Firenze, Italy

GIANPIETRO DEL PIERO
  Dipartimento di Ingegneria, Università di Ferrara
  via Saragat 1
  I-44100 Ferrara, Italia

IRENE FONSECA
  Dept. of Mathematical Sciences, Carnegie Mellon University
  Pittsburgh, PA 15213, USA

ILARIA FRAGALÁ
Dipartimento di Matematica "F. Brioschi"
Politecnico di Milano
Piazza Leonardo da Vinci 32
I-20133 Milano, Italy

NICOLA FUSCO
Dipartimento di Matematica e Applicazioni
Università di Napoli
Via Cintia
I-80126 Napoli, Italy

ANTONIO LEACI
Dipartimento di Matematica "Ennio De Giorgi"
Università di Lecce
Via Arnesano
I-73100 Lecce, Italia

GIOVANNI LEONI
Dipartimento di Scienze e Tecnologie Avanzate
Università del Piemonte Orientale
I- 15100 Alessandria, Italy

RICCARDO MARCH
Istituto per le Applicazioni del Calcolo, CNR
Viale del Policlinico 137
I-00161 Roma, Italy

JEAN MICHEL MOREL
CMLA, ENS Cachan,61, Av. du Président Wilson
F-94235 Cachan Cedex France

EDOUARD OUDET
Département de Mathématiques
Université Louis Pasteur
7 rue René Descartes
F-67000 Strasbourg, France

MAURIZIO PAOLINI
Dipartimento di Matematica e Fisica
Università Cattolica di Brescia
I-25121 Brescia, Italy

FRANCO PASQUARELLI
Dipartimento di Matematica e Fisica
Università Cattolica di Brescia
I-25121 Brescia, Italy

DANILO PERCIVALE
Dipartmento di Metodi e Modelli Matematici
Università di Genova
P.le J.F. Kennedy Pad. D
I-16129 Genova, Italia

JAYANT SHAH
> Mathematics Department, Northeastern University
> Boston, MA, USA

EUGENE STEPANOV
> Dipartimento di Matematica, Università di Pisa
> via Buonarroti 2
> I-56127 Pisa, Italy

FRANCO TOMARELLI
> Dipartimento di Matematica "Francesco Brioschi"
> Politecnico di Milano
> Piazza Leonardo da Vinci 32
> I-20133, Milano, Italia

List of Participants

Emilio Acerbi (Parma)
Micol Amar (Pavia)
Alberto Bersani (Roma)
Primo Brandi (Perugia)
Giuseppe Buttazzo (Pisa)
Valeria Chiadó Piat (Torino)
Fabrizio Colombo (Milano)
Sergio Conti (Leipzig)
Giampietro Del Piero (Ferrara)
Selim Esedoglu (Minneapolis)
Irene Fonseca (Pittsburgh)
Nicola Fusco (Firenze)
Adriana Garroni (Roma)
Maurizio Grasselli (Milano)
Enrico Laeng (Milano)
Antonio Leaci (Lecce)
Francesco Leonetti (L'Aquila)
Roberto Lucchetti (Como)
Clelia Marchionna (Milano)
José Matias (Lisbona)
Jean Michel Morel (Paris)
Stefano Mortola (Milano)
Tullia Norando (Milano)
Maurizio Paolini (Brescia)
Danilo Percivale (Genova)
Severine Rigot (Orsay)
Ernesto Salinelli (Novara)
Rosanna Schianchi (Roma)
Francesca Sianesi (Milano)
Margherita Solci (Pavia)
Franco Tomarelli (Milano)
Enrico Vitali (Pavia)

Giovanni Alberti (Pisa)
Francesca Antoci (Novara)
Guy Bouchitté (Toulon)
Ariela Briani (Pisa)
Antonin Chambolle (Paris)
Claudio Citrini (Milano)
Giuseppe Congedo (Lecce)
Gianni Dal Maso (Trieste)
Francois Ebobisse Bilie (Trieste)
Giorgio Follo (Milano)
Ilaria Fragalá (Milano)
Emma Gallo (Pavia)
Alessandro Giacomini (Trieste)
Robert V. Kohn (New York)
Maria Rosaria Lancia (Roma)
Gian Paolo Leonardi (Povo-Trento)
Giovanni Leoni (Alessandria)
Riccardo March (Roma)
Silvia Mataloni (Roma)
Giuseppe Mingione (Parma)
Massimiliano Morini (Trieste)
Umberto Mosco (Roma)
David Owen (Pittsburgh)
Eduardo Pascali (Lecce)
Marcello Ponsiglione (Trieste)
Irene Maria Sabadini (Milano)
Anna Salvadori (Perugia)
Jayant Shah (Boston)
Florian Sobieczky (Berlino)
Mikahil Sytchev (Leipzig)
Lev Truskinovski (Minneapolis)
Elvira Zappale (Salerno)

Progress in Nonlinear Differential Equations and Their Applications

Edited by
Brezis, H., Université de Paris, France / Rutgers University, USA

Progress in Nonlinear Differential Equations and Their Applications is a book series that lies at the interface of pure and applied mathematics. Many differential equations are motivated by problems arising in diversified fields such as mechanics, physics, differential geometry, engineering, control theory, biology and economics. This series is open to both the theoretical and applied aspects, hopefully stimulating a fruitful interaction between the two sides. It will publish monographs, polished notes arising from lectures and seminars, graduate level texts, and proceedings of focused and refereed conferences.

Vol. 50
Lorenzi, A. / Ruf, B., both Università degli Studi di Milano, Italy (Eds.)

Evolution Equations, Semigroups and Functional Analysis
In Memory of Brunello Terreni

2002. 412 pages. Hardcover
ISBN 3-7643-6791-1

Brunello Terreni (1953-2000) was a researcher and teacher with vision and dedication.
The present volume is dedicated to the memory of Brunello Terreni. His mathematical interests are reflected in 20 expository articles by distinguished mathematicians. The unifying theme of the articles is evolution equations and functional analysis, which is presented in various and diverse forms: parabolic equations, semigroups, stochastic evolution, optimal control, existence, uniqueness and regularity of solutions, inverse problems as well as applications.

Vol. 49
Benci, V., Universita di Pisa, Italy / **Cerami, G.**, Universita di Palermo, Italy / **Degiovanni, M.**, Universita Cattolica del Sacro Cuore, Brescia, Italy / **Fortunato, D.**, Universita di Bari, Italy / **Giannoni, F.**, Universita di Camerino, Italy / **Micheletti, A. M.**, Universita die Pisa, Italy (Eds.)

Variational and Topological Methods in the Study of Nonlinear Phenomena

2002. 136 pages. Hardcover
ISBN 0-8176-4278-1

This volume covers recent advances in the field of nonlinear functional analysis and its applications to nonlinear partial and ordinary differential equations, with particular emphasis on variational and topological methods.

For orders originating from all over the world except USA and Canada:
Birkhäuser Verlag AG
c/o Springer GmbH & Co.
Haberstrasse 7, D-69126 Heidelberg
Fax: ++49 / 6221 / 345 229
e-mail: birkhauser@springer.de

For orders originating from USA and Canada:
Birkhäuser
333 Meadowland Parkway
USA-Secaurus, NJ 07094-2491
Fax: ++1 / 201 / 348 4033
e-mail: orders@birkhauser.com

Birkhäuser